THE RIGHT TO OBLIVION

THE RIGHT
TO OBLIVION

Privacy and the Good Life

LOWRY PRESSLY

Harvard University Press

CAMBRIDGE, MASSACHUSETTS   LONDON, ENGLAND   2024

Library of Congress Cataloging-in-Publication Data
Names: Pressly, Lowry, author.
Title: The right to oblivion : privacy and the good life / Lowry Pressly.
Description: Cambridge, Massachusetts ; London, England : Harvard University Press, 2024.
  | Includes bibliographical references and index.
Identifiers: LCCN 2024001368 | ISBN 9780674260528 (cloth)
Subjects: LCSH: Privacy. | Quality of life. | Self. | Information society—Moral and ethical
  aspects.
Classification: LCC GT2405 .P74 2024 | DDC 323.44/8—dc23/eng/20240324
LC record available at https://lccn.loc.gov/2024001368

For you know who

One would have wanted more—more—more—
Some true interior to which to return,
A home against one's self, a darkness,

An ease in which to live a moment's life,
The moment of life's love and fortune,
Free from everything else, free above all from thought.

—*Wallace Stevens*

The subject of this book is not the void exactly, but rather what there is round about or inside it.

—*Georges Perec*

Contents

Introduction   1

1 Photography and the Invention of Privacy   23

2 Privacy, Perception, and Agency   58

3 Hiding in Private   88

4 Memory and Oblivion   118

5 Privacy and the Production of Human Depth   148

   Postscript   177

   Notes   181

   Acknowledgments   215

   Index   217

THE RIGHT TO OBLIVION

"Neighbors #58." Courtesy of Arne Svenson and Robert Klein Gallery.

INTRODUCTION

We live in an age blinded by information. Endless streams of updates from friends, family, and celebrities find us wherever we are, in between notifications of far-flung disasters and videos of strangers ambling down the streets and alleyways of cities we will never visit. The recorded facts of science and history are constantly at our fingertips. We have precise data about the beating of our hearts, the number of steps we take in a day and where they take us, the fluctuation of the temperature in our homes, what we do in our sleep. We know more about ourselves, others, and the world than ever before. Does this unprecedented bounty of information about human life mean that we see it more clearly? More deeply? I am not so sure.

What is especially impaired by information in our time is our faculty of critical and moral vision, that faculty by which we *see problems*. In the digital age, we have come to view a great deal of human life, both what we know of it and what we do not, through the lens of information. Conversation is an exchange of it. Our intimates are those with whom we share the information we withhold from others. People are described as "data subjects."[1] "You are your data" is how sociologist Deborah Lupton put it in her study of the quantified-self movement, whose adherents seek to get better in touch with their lives by generating as much information about themselves as possible.[2] The assumption that the more we know, the better, is so profoundly embedded in our culture that it is hard even to notice it. Yet in the welter of this increase, we have lost sight of valuable regions of life that depend upon the preservation of limits to the knowable.

It is counterintuitive to suppose that our ability to perceive and inhabit the world, to know ourselves and one another, can be undermined by knowledge and not just supported by it. You cannot inhabit the world in a state of total obliviousness, only drift through it. You cannot love someone you do not know. Democratic self-governance and collective organization require a great deal of information to be effective. Saying that we are *blinded by information* is a bit like saying that we could be *impoverished by money*. These phrases share more than an air of apparent contradiction, however: they also express a similar moral idea. Some people really are impoverished by money, not in the sense that by having more they somehow have less of it, but rather that their lives go less well as a whole because they have come to view too much of human life in terms of dollars and cents. The same holds true for knowledge and information. This is what this book aims to demonstrate: that alongside the obvious benefits of knowledge, we also need limits to what it is possible to know, and limits of a particular sort. Without such limits and the realms of unknowable experience they produce, human life is significantly impoverished, shallower, and less human.

Philosophers do not tend to argue for the importance of knowing less. The motto of the profession is still largely Kant's *sapere aude* ("dare to know!"), itself a revision of the Delphic "know yourself" for an age of Enlightenment captivated by exploration and the will. The famous encomium to self-knowledge carved into the Apollonian temple at Delphi is still our touchstone for the value of wisdom, and with good reason. But we should not forget that the commandment to know oneself was accompanied by another, etched into the very same temple stone: "Nothing in excess." The implication is that there are ethical limits to self-knowledge, which are based, as the second maxim indicates, in a broader view of human well-being that includes knowledge but also other values seemingly opposed to it and necessitating its moderation. This is an idea at odds not only with the form of rational inquiry into human affairs practiced by the humanities and social sciences, but also with today's information-based political economy and culture of identity, authenticity, and self-branding. Nonetheless, the thought is not entirely lost to us. It still appears in the statements and work of artists, for instance, who, unlike the scholar and middle manager, are free to employ omission and opacity in order to lend their works a sense of depth, liveliness, and resistance to exhaustive interpretation. Later in this book, and especially in the conclud-ing chapter, I will have much more to say about what it means for an individ-ual life and the shared world of collective activity to have this sort of depth and how it is produced by privacy's particular limits to knowledge. So for now, consider a provocative expression of the idea.

In the midst of a critique of modern "psychology and self-reflection," the filmmaker Werner Herzog echoes the arguments I will make about priva-cy and the social production of oblivion: "Explaining and scrutinizing the human soul, [peering] into all its niches and crooks and abysses and dark corners, is not doing good to humans. We have to have our dark corners and the unexplained. We will become uninhabitable in a way an apartment will become uninhabitable if you illuminate every single dark corner and under the table and wherever—you cannot live in a house like this anymore. And you cannot live with a person anymore—let's say in a marriage or a deep friendship—if everything is illuminated, explained, and put out on the table."[3] This is very close to my own answer to the question of what privacy is and why it is valuable. However, for the human soul—or person, let us say—and the world in which it dwells, the ways of obscurity are many. Not all dark corners are alike. Things can be secret, hidden, lost, anonymous, unnoticed, private, obfuscated, opaque, translucent, and a great deal more. How we understand these limits to knowledge and what lies behind them has significant consequences for how we relate and respond to such lim-itations, and for the sort of person, relationships, and world they produce. The habitability of the self and the world requires, I will argue, a particular kind of not-knowing, which I call *oblivion* to highlight its most essential and distinctive qualities. Oblivion describes a form of the unknown that, unlike what is secret or hidden, is essentially resistant to articulation and discovery: its limits do not concern who knows what, but instead what can be known. The preservation of oblivion in our world and ourselves is essen-tial for the sense of potentiality, depth, play, and freedom in human affairs. Privacy is one way that we integrate oblivion into our daily lives, although as we will see, there are others, too.

The Ideology of Information

Contemporary debates over privacy are blinkered by an unreflective tenden-cy to understand everything in terms of "the information age," under whose ideological sway so much of contemporary life takes place. I do not wish to deny the importance of paying attention to the vast quantities of personal information that we are constantly generating. Surely we are correct to think that we have, or ought to have, moral and legal rights to exercise control over such information and to protect us from the harms that can ensue when it falls into the wrong hands. But we are mistaken to think that this is what *privacy* is for. One of the main claims of this book is that privacy is valuable not because it empowers us to exercise control over our information, but because it protects against the creation of such information in the first place. This is not a technophobic or reactionary crusade against the existence of information as such, but rather an argument that individual and collective well-being require regions of possibility and experience that are essentially opposed to the fixity of information, and that the value of privacy and related practices, rights, and norms derives from how they instantiate these regions in our everyday lives. Such an argument will naturally have its opponents. The most obvious of these are the myriad projects of state, corporate, and citizen surveillance that seek to convert as much of our lives as possible into information so that it may be commoditized and sold, used to influence our behavior, or otherwise stored for future use. Yet this argument has another, perhaps unexpected opponent, whose threat to privacy is also quite serious, albeit much less obvious. That opponent is the standard understanding and defense of privacy in our time, which, by accepting the ideology of the information age, both misses privacy's true value and unwittingly aids the forces it takes itself to be resisting.

Bernard Harcourt has noticed a change in how we talk and think about privacy over the period of history corresponding to the rise of neoliberalism. The marked shift from discussing privacy's "virtue or even its value" to debating its "costs and benefits" is, he claims, the result of a broader trend of thinking about political and ethical life in predominately economic and market terms.[4] To be sure, the neoliberal wave had all sorts of consequences for ideas about humanity and its well-being. Market-oriented values of liberty, choice, and property in identity and culture have for decades been ascendent in both theoretical and commonsense conceptions about what makes a life go well and what we owe to one another. The consequences of this shift are so numerous and well-documented that they have by now become a kind of cliché, with neoliberalism a ready-made explanation for all of society's ills. Nevertheless, Harcourt was among the first to notice its effect on our thinking about privacy. The dominant view of privacy today is of something like a personal preference, à la social media "privacy settings" and the interminable "privacy policies" we must accept before using a particular service, whose value may vary from person to person depending on how much they care about it. "Whereas privacy was previously framed in humanistic terms, it is now far more likely to be thought of as a type of property, something that can be bought and sold in a market."[5] Privacy on this view is good if and when we want it, which makes deciding to exchange it for other things we want and at a value set by the market unproblematic. Hence the overwhelming focus on control, consent, and access in both contemporary defenses and deprivations of privacy.

Harcourt is right about this trend in thinking about privacy, which is decidedly at odds and still in tension with a view of privacy as a fundamental human right. The idea that the "services" we receive from social media and fitness trackers are given to us in exchange for our information is one omnipresent example. So too is the idea that we voluntarily part with privacy by going out in public, from which it is thought to follow that we cannot reasonably expect to have privacy rights and interests when out among others. We want a paycheck or a diploma, so we tacitly agree to pay for it by having our keystrokes or eye movements surveilled. One critique of these practices draws on the moral language of coercion and argues that the stark asymmetry of power behind these relationships precludes any idea of consenting to them. This is true, but it is not an argument about surveillance or privacy as such. What's more, the counterfactual, consensual privacy transaction only appears morally unproblematic if we think that privacy is one fungible good among others, which may be exchanged for other things we might want. The idea that privacy can be traded like any other form of personal property is an idea of our time, reflective of the particular arrangements of political and economic power in the era of the data economy and surveillance capitalism.

But there is another ideology critique to be made about privacy, based not in the ideology of neoliberalism but in that of the information age, and which is aimed not just against the surveillance industry or even those who treat privacy as a mere preference, but rather at the culture as a whole, including privacy's theorists and defenders. In addition to focusing on the effects of market ideology under neoliberalism, we should look also to the various pressures that encourage us to understand as much as possible about ourselves and others in terms of information. I will refer to the effect of this broad range of developments as "the ideology of information" in order to pick out an important but unspoken premise underlying much contemporary discourse around privacy and the ethics of technology. This is the idea that information has a natural existence in human affairs, and that there are no aspects of human life which cannot be translated somehow into data.

One is tempted to recite the Marxian aphorism about the ruling ideas of a time being the ideas of its ruling class, for the assumption that everything can be unproblematically understood or even represented by aggregated information is obviously an idea in the interest of surveillance states and capitalists.[6] It marks the boundaries of resource extraction at the very limits of human existence. When it comes to contemporary ideas about the importance and scope of privacy, one consequence of this ideology is that the existence of information in the private realms of human life is taken for granted, naturalized, and treated as prior to the question of what privacy does or is for. Debates over the value of privacy are then restricted to questions about whether its value has to do with its usefulness for controlling, protecting, or restricting access to some information that already exists. Political and ethical questions about the existence of such information in the first place are rendered invisible or transmuted into different and less fundamental concerns about the consequences that can flow from information's misuse or misplacement.

Pick up a book or article on privacy written in the past twenty years and you will likely find an example of the ideology of information at work. So ubiquitous is the assumption that privacy describes some form of protection, concealment, or control of information that giving examples seems arbitrary, if not gratuitous. Consider this description of the field of thought from the *Stanford Encyclopedia of Philosophy*, surely the most thorough and authoritative compendium of contemporary thinking on the subject. The entry repeats the idea, common among scholars, that there are, in fact, two kinds of privacy. The first of these it calls "constitutional (or decisional) privacy," which in fact does not refer to privacy at all but to other values like autonomy, negative liberty, and bodily integrity that courts, especially in the United States, have smuggled into the law under the cloak of privacy jurisprudence.[7] The connection of these other values with privacy owes partly to decades of creative jurisprudence and activism, and partly to the liberal tradition in political thought which advocates for the existence of a "private sphere" of human activity beyond the legitimate reach of state influence in citizens' lives. The value of this sphere doesn't have anything to do with the value of *privacy*, but instead with the moral importance of letting people determine the course of their own lives.

The ordinary meaning of privacy is epistemic, having in some way to do with limits to perception and knowledge. It is what we have or want when we are in bathroom with the door closed. The encyclopedists describe *that* kind of privacy as "concerned with the interest of individuals in exercising control over access to information about themselves and is most often referred to as 'informational privacy.' Think here, for instance, about information disclosed on Facebook or other social media. All too easily, such information might be beyond the control of the individual."[8] The idea expressed here is so common today that it is practically universal: that all privacy in the ordinary sense of the word is properly called "informational privacy" because it is concerned with the protection, control, access, or "contextual integrity" of personal information, which "flows" from one side of privacy's barrier to the other.[9]

This is an understanding of privacy plagued by deep conceptual incoherencies, which I will draw out at length over the course of this book. So at this point let us notice that as a *defense* of privacy, it is also especially hamstrung. The idea is not just impoverished but dangerous because it naturalizes the existence of information prior to the question of privacy. It takes for granted that some information will always already preexist moral and political questions about privacy, which are then merely concerned with controlling or protecting it. As one privacy scholar puts it, "Informational privacy is thus recast in terms of the protection of personal data."[10] This is the view of privacy that one gets from interacting with so-called privacy settings online, where one can make one's posts on social media or the biometric data from one's smart watch "private," meaning that one may prevent other *users* and entities beyond the platform from accessing it or taking it out of context. But by then, it is already too late. The surveillance states and capitalists who own and control the platforms on which we might exercise the limited choices of "privacy settings" already have what they need once some piece of information about us is created, after which it matters little to them with whom we choose to share it. Seen from this point of view, the eagerness with which Internet companies always tout their latest privacy features appears less like a commitment to protecting human interests at the cost of profit maximization than cheerleading for a diminished value of privacy in service of their bottom line.

There is nothing inherently wrong with using the technologies of our time to think about the political and ethical dimensions of personal knowledge, although we should be skeptical about using concepts and metaphors derived from Internet technologies—privacy as controlling flows of information, for example—insofar as those technologies are the profit-maximizing engines of an industry that depends upon the production and harvesting of personal information.[11] The encyclopedists' choice of example to explain what they mean by privacy is extremely revealing, if entirely common, in this regard: "Think here, for instance, about information disclosed on Facebook or other social media." By the same token, we should hesitate before interpreting Mark Zuckerberg's famous quip about privacy no longer being a social norm as a sincere appraisal of the social relevance of privacy instead of a wishful expression of his company's business model. But of course the technologies of the digital age are not the only ones that have been used to come to grips with privacy's value.

The historical portions of this book will tell the story of how a different and, I will argue, preferable view of privacy emerged in reaction to the invention of the photograph camera and the rise of mass media in the nineteenth century. Those earlier arguments opposed a humanistic moral rhetoric to what they perceived as anti-human threats to privacy, whereas defenses of privacy in the information age largely accept the terms of their opponents before the argument even begins. Any view of privacy that takes for granted the existence of some information that is controlled, hidden, and shared by the person to whom it pertains is not just incapable of opposing the arrangements of power associated with corporate, state, and citizen surveillance. Such views also cede vital political ground to their opponents by naturalizing the existence of personal information when instead they should be, as they say, problematizing it.[12]

Yet the problem is more serious than this, for as we will see, there are vital parts of human life that are inimical to being turned into information. Privacy, writ large, does not just protect these unaccountable regions of life, in which there is no fact of the matter about us, but is necessary for their very existence. Proceeding in tandem with this book's historical excavation is a philosophical argument that aims to show how such regions of oblivion in individual and collective life are necessary for the sense that our lives are up to us and worth the trouble of living, that individuals and the social world of shared experience have depth and meaning, that one can trust oneself and others, and much else. Privacy's regions of oblivion are not zones of pure ignorance, absence or void: they can be experienced, encountered, and respected in modes other than knowledge and control.[13] It is only thus that privacy can offer, in the words of the epigraph from Wallace Stevens, "Some true interior to which to return, / A home against one's self, a darkness, / An ease in which to live a moment's life." By circumscribing the role of knowledge and information in human affairs, privacy permits us to become acquainted with the ambiguous parts of ourselves by being ambiguous ourselves and to enjoy the range of human experience that lies beyond the limits of knowledge and control.

So far I have quoted a poet and a filmmaker, but have only made passing reference to a Silicon Valley CEO. It would be fair to wonder whether I am offering an eccentric or romantic view of privacy, nice to think about but divorced from the more quotidian role that concealment and exposure play in structuring life in the digital age. But that could not be further from the truth. Behind this objection lies something like the distinction Joshua Rothman draws between two conceptions of privacy: the "citizen's sense of privacy," on the one hand; and the "artist's sense of privacy," on the other. According to Rothman, those who take the citizen's view concern themselves with "other people and how they might affect us . . . how they could use information about us for their own ends, or interfere with decisions that are rightfully ours." Rothman contrasts this perspective with those who, like Virginia Woolf, take the artist's view and see privacy as having "something to do with preserving life's mystery; with leaving certain things undescribed, unspecified, and unknown" which rests upon "an intensified sense of life's preciousness and fragility, and on a Heisenberg-like notion that, when it comes to our most abstract and spiritual intuitions, looking too closely changes what we feel. It has to do, in other words, with a kind of inner privacy, by means of which you shield yourself not just from others' prying eyes, but from your own."[14]

Yet it turns out that these views are far less distinct than Rothman supposes, both in substance and point of origin. By looking through the history of the public moral and political discourse around privacy, we will see that the civic aspects of privacy have, to a significant extent and from the very beginning, been animated by what Rothman calls "the artist's sense." Since the earliest calls for rights to privacy in the nineteenth century, citizens have advocated a view of privacy as preserving a condition of potentiality, flux, and play in which the life of the citizen and the life of the artist are most alike. Conversely, we will also see that what Rothman calls the artist's sense of privacy is not confined to giants of the imagination like Virginia Woolf, but is in fact a crucial and animating element of our ordinary values and moral intuitions about the importance of privacy in human life.

This account may be untimely in an age of information, but it is by no means brand new. A sense that privacy is fundamentally opposed to information has animated public moral discourse on the subject since the very beginning. Indeed, the history of privacy advocacy up until the last several decades reveals such a constant concern with the incursion of information into the unarticulated realms of human experience that we ought to consider the age of information and the age of privacy anxiety to be more or less coextensive. It is highly significant that the period in which human life began to be documented in ever increasing detail and quantity—roughly around the middle of the nineteenth century, when photography, newspapers, and other mass media, Bertillionian biometrics, and many other cognate developments appeared on the scene with increasing rapidity—was also the period in which articles began to appear with titles like "Is There Any Privacy?," "The Decay of Privacy," and "Is the End of Privacy Coming to Human Kind?"[15] The lament over privacy's imminent demise is a consistent refrain and distinctive marker of the long information age that stretches from the nineteenth century to our own era overflowing with breathless writers proclaiming "The End of Privacy" at the hands of the latest technical innovation.[16] What is old is new again, and what is new, old. This is not to say that contemporary threats to privacy are not serious—quite the opposite—only that we should be careful not to ignore the context in which they appear, for to do so is to risk misunderstanding the nature of the threat and deprive ourselves of insights from a long tradition of thinking on the subject.

While many technologies of the digital age do pose novel ethical and political challenges—mobile connectivity and the Internet being two of them, discussed in chapters 3 and 4, respectively—the basic moral frameworks for understanding the dangers of information go all the way back to the nineteenth century. Their concerns were, by and large, our concerns, although by abandoning their critical stance to the existence of information in human affairs, we are less prepared to understand and meet them in our time. Indeed, when we set our contemporary moment in its larger historical context, and when we abandon the myopic presentism of the technology industry (whose ceaseless sloganeering of "innovation" and "revolutionary technologies" reflects market imperatives more than any historical reality), we see that the most novel threat of our time might not be any particular piece of technology, but the widespread acceptance of the ideology of information.[17]

The moral language of privacy has been the common idiom for citizens of the nineteenth, twentieth, and twenty-first centuries to express their anxieties about the political economy of information and its raids on the unarticulated regions of human experience. For this reason, the invocation of privacy against these anxieties frequently extends beyond what is recognizably private into situations where information may be said to be divulged with consent or obtained from the public sphere—as with, for instance, photographs taken in public, the journalism at the center of debates over the right to be forgotten, and concerns about being surveilled, tracked, and subject to facial-recognition technology. This stretching of its concept presents a problem for privacy while also reiterating the force of its moral idiom. For although an intuitive commonality gathers these various cases together, the conceptual language of privacy is ill fitted to capture the continuity between them, especially when it comes to the ethical dimensions of knowledge in public.

The history of privacy advocacy suggests a way to make sense of these intuitions without having to insist on interests of "privacy in public," whose air of paradox tends to undermine both privacy and whatever associated interests we have when we step out of doors. In chapter 1, for instance, we will see how the reaction of early privacy advocates to the invention of photography expressed moral outrage and anxiety about the conversion of the fluidity of everyday life into the fixity of information, no matter whether at home or on the street. Chapter 3 likewise traces complaints about the quantity and quality of publicity that individuals willingly invite into their lives, from the spread of newspapers in the nineteenth century, and radios and television in the twentieth, to today's era of mobile connectivity and social media. Chapter 4 recounts movements against the massive accumulation of information about individuals, from databanks to the Internet and the right to be forgotten. By interweaving this history with philosophical analysis, we will see that although it is a mistake to think that the panoply of critiques of information popularly associated with privacy really do all concern privacy as such, the common intuition linking them nevertheless expresses a truth: they are all animated, in one way or another, by the value of oblivion.

On Oblivion

There are many ways not to know something. We can be ignorant, baffled, oblivious; obdurate, obtuse, suspicious; insensate, confused, deceived by ourselves and others. We forget. We confront mysteries and suspect others of keeping secrets from us. Much in this world is hard to understand. We speak differently about these various forms of not-knowing because they are meaningfully different. For instance, suppose I am oblivious to the fact that I am being recorded by a hidden camera, which you have placed in the knot of a tree in a public park. From your perspective, my obliviousness about the camera will look different than if I were mystified by it (wondering, say, at my inverted face in the convex glass of the lens), or if I suspected but could not confirm its existence. The experience of these various modes of unknowing is different for me, too. To be willfully ignorant is not the same as to be unlearned, oblivious, suspicious, searching for the hidden, or baffled by a mystery. Although the information I lack is the same in every case—that is, I am unaware of the camera's existence—my experience of that lack is not, and this difference affects my relation both to the world and to myself. It is significant for my marriage whether I regard that portion of my wife's life to which I am not privy—say, the text messages on her phone or what she is thinking in a pensive moment—as private, secret, hidden from me, or mysterious. The same goes for the stance I assume toward myself, my thoughts, my past, my potential to be different.

If the forms of not knowing are manifold, then how much more various must be the ways that we are unknown to one another and ourselves? We can be coy, ambiguous, and ironic; reticent, secretive, deceptive, and unintelligible. Who hasn't been each of these and more at one time or another? We seek privacy, go into hiding, are confined to the closet, disappear in the anonymous crowd. We are forgotten. We can, like Ralph Ellison's invisible man, pass unseen and unrecognized through a society blinded by ideology and self-interest: "I am invisible, understand, simply because people refuse to see me. Like the bodiless heads you see sometimes in circus sideshows, it is as though I have been surrounded by mirrors of hard, distorting glass. When they approach me they see only my surroundings, themselves, or figments of their imagination—indeed, everything and anything except me."[18] The regular attendees of an Alcoholics Anonymous meeting are unknown to one another in a different way than are strangers on the street or the nametag-wearing denizens of professional conferences and hospitals. There are no secrets in a marriage, so they say, and whether this is true or not it is undeniable that the daily life of even the most forthcoming couple is structured from morning to night by privacies. Structure the same marriage as thoroughly with secrets, however, and it falls apart: trust turns to suspicion, and the respect for the limits of one's knowledge gives way to the grasping impatience of surveillance.

This book begins from the assumption, based in pervasive usage and common sense, that this wonderfully multifarious language for describing the ways of opacity reflects a real variety in the modes of concealment and obscurity; that these differences are not only epistemic, but also ethical, political, and existential; and that, as the examples above make clear, they do not identify primarily natural (biological, optical, etc.) limitations to knowledge, but human-made ones. I do not merely assume this but will argue for it at length, because the elision of privacy with other forms of concealment or opacity—secrecy, anonymity, confidence, hiding—is so common as to be practically foundational to contemporary debates on the matter. It is, at the very least, a major stumbling block, although as with the ideology of information, the danger is not merely conceptual, but also political and ethical.

The reality of privacy, as distinct from the reality of physical walls or being unseen or unheard, is the reality of a social phenomenon. It consists largely in how we understand and value it, and what practices, norms, laws, and so forth we develop on the basis of that understanding. As a set of practices and values, privacy can be diminished, corrupted, or changed simply by our forgetting or reimagining what its boundaries protect and what lies beyond them. If the role and value of privacy in our lives is not the same as that of secrecy, for instance, then a failure to note the difference between them puts us at risk of making mistakes when we need to determine the proper scope of privacy or weigh it against the other important interests with which it will sometimes be in tension.

If privacy is something like a fundamental human interest or right, whereas the value of secrecy is largely contingent upon what is kept secret (surprise party plans, insider trading), then by confusing the difference between the two we run the risk of one day restricting something fundamental to well-being when we think we are giving up something far less valuable. This is the malevolent intent of the propagandistic claim that "privacy is only of value for those who have something to hide." That a slogan this fatuous could survive for so long is a testament to our failure to be clear about distinguishing privacy from secrets, hiding, and the rest. Our muddied understanding forces us into lame replies like "everybody has something to hide" rather than saying what we ought to say: "Hiding is for those with something to hide!" Secrecy is for keeping secrets. Privacy is for something else. The idea that privacy is for controlling access to information or my person makes this distinction seem unavailable or indeed unthinkable.

Consider the difference between privacy and secrecy. My five-year old daughter mentioned that she had something planned for my upcoming birthday. "What is it?" I asked. "It's a secret," she said. Very well. However, had she said, "That's private," I would have been taken aback; maybe I would have laughed and chalked it up to her being a child still learning the rules of our language. The distinction is even more notable in the gulf between a friend's "private life" and "secret life," or in the perversity of calling an adulterer's "secret family" his "private family." Secrecy does what the standard informational view thinks that privacy is for: it protects certain information, permits an individual to control who has access to it, and gives us a moral language for describing the disrespect and violation of that control or access.

For something to be a secret, it must be known, and although the secret typically takes the form of a piece of information, we also speak of secret fishing holes or lovers' secret hideaways.[19] In any case, a secret must be known definitively and unambiguously by at least one person for it to be a secret. If you tell me, and only me, your secret recipe for fried chicken, and I forget it or mix it up, then I am no longer in possession of your secret. If you forget it, too, then there's no longer any secret to speak of. Secrets can be shared, and how they are shared is that someone else comes to know them, again definitively and without ambiguity. If I only tell you half my secret, or my description of it is ambiguous or contradictory, then you do not share it. If you do come to know my secret—either by my revealing it or your discovering it (note the connection to hiding as well as the implicit assumption that there is something there to be uncovered)—my secrecy is not thereby destroyed, but you are now "in on it." This fact about the durability of secrets is the coercive leverage of blackmail and the narrative engine of many a thriller: "If you want this to stay a secret then you had better do as I say." This is obviously different from privacy, which is not durable but is "destroyed" or "violated" when it is breached. The Peeping Tom at your bathroom window cannot share or "be in on" your privacy the same way that he would share a secret of yours if he discovered it but told nobody.

So, if privacy does not protect secrets, what does it do? My answer is that it protects, but also produces, oblivion. I use the language of oblivion to refer to a form of obscurity that does not conceal some information but describes a state of affairs about which there is no information or knowledge one way or the other, only ambiguity and potential. These qualities of oblivion are destroyed when translated into information and other forms of knowledge that require definition and noncontradiction. By contrast to the secret, which in its concealment must be definite to be a secret at all, there is no fact of the matter about what is concealed by oblivion, no way to say that it is either x or not x. This is why telling a lie in response to an invasive question about your sex life might preserve your secret but not your privacy. By its very impenetrability and resistance to knowledge, oblivion lends individual agency and self-relation, as well as the collective life of sociality and politics, irreplaceable support for the sense that life, both individual and collective, has depth as well as the possibility for change and surprise. For oblivion to contribute to the depth and meaningfulness of human life, it cannot simply be a void or absence, like the far cold reaches of deep space or the unknowable time before the big bang, but must be capable of being integrated and encountered in ordinary human affairs. This is what privacy does, why it is valuable and worth defending.

My use of the concept of oblivion draws upon the root meaning of the term as what is lost from memory, but also the mental and social state of "obliviousness" as a particular mode of unknowing, as well as oblivion's connotation of impenetrability to human powers of perception, knowledge, and control. The word derives from the Latin for forgetting (*oblivio*), whose prefix *ob-* (meaning toward, on, or against) already indicates that although oblivion is a kind of absence, it is nevertheless something that can be confronted, that can constitute a form of presence and positive reality. This original sense persists in the Romance language words for forgetting (*olvidar, oublier*), which is what in those languages we are said to have a right to when we have a right to anonymize or delete personal information on the web. The central idea of mnemonic oblivion is that of a barrier to knowledge that does not conceal some discrete piece of information or memory but marks the limit beyond which even the greatest human powers of perception and understanding cannot pass. As Buckingham says to Richard III, it is "the swallowing gulf/ Of dark forgetfulness and deep oblivion."[20] Shakespeare's line evokes the other most common association of oblivion with limits to perception and knowledge: the barrier erected by death. As in death, what is forgotten does not exist somewhere in one's mind or beyond it, waiting to be remembered, but is gone beyond recall. The English phrase "the right to be forgotten" likewise draws upon this total quality of oblivion. To be forgotten means that the world is oblivious to your existence.

We also call certain people "oblivious" to describe a particular way of relating to themselves and to the world of others. The usage is often pejorative. The oblivious person is *clueless*, someone who should have noticed something but has not. Yet our judgment is not always so negative. If I am oblivious to my surroundings, I might be worse at navigating through them, but I might also be wildly happy, lost in reverie, or deep in thought. The point is that obliviousness describes a particular form of not knowing: One doesn't know what one doesn't know and, just as important, does not suspect that there is something unknown to them, hidden away. This is just the sort of unknowing that we want from the world at large with regard to our private lives: we don't want people staring at our closed drapes or listening under our eaves, but to mind their own business without concerning themselves about what we might or might not be concealing. This is what we mean by the phrase "keep your nose out"—where the nose points, there go the eyes and attention. When we respond to an invasive question by saying "that's private," what we are trying to do is maintain the obliviousness of the one who asked it, which we risk losing if we respond to the question by blushing or by saying "it's a secret." Again, this is why telling a lie in response to an invasive question can preserve one's secret but not one's privacy: the lie creates a piece of information, albeit a false one, where before there had been none. A secret can survive this change, but oblivion cannot.

The past- and present-oriented senses of oblivion come together in the idea, at least as old as Plato, that knowledge requires both memory and perception. If oblivion is a form of not-knowing, then we should expect it to have both perceptual and mnemonic dimensions. The two senses commingle historically in one of the most longstanding uses of the term to refer to laws of amnesty passed in the aftermath of civil conflict, so-called Acts of Oblivion. The Treaty of Westphalia marked the end of the Thirty Years' War in 1648 by declaring that "There shall be . . . a perpetual Oblivion . . . of all that has been committed since the beginning of these Troubles . . . [All] shall be bury'd in eternal Oblivion."[21] Acts of Oblivion are meant to put a stop to cycles of violent reprisal by outlawing the bearing of grudges, or at least acting on them. But they also prescribe a social ethic of deliberate obliviousness, either in the spirit of "all shall be bury'd" or by explicit prohibition of speech and writing on the subject.[22] Obviously, nobody who lived through thirty years of war was going to forget about it, nor have we, some four centuries later. Rather, Acts of Oblivion seek to enforce a norm of interpersonal comportment that leaves certain matters unarticulated in human affairs. This is something that rights and norms of privacy aim at, too, which not only prevent us from knowing certain things about others but also demand that we respect others' privacy. Sometimes this will mean not snooping or invading someone's space, while other times it will mean maintaining an attitude of willed obliviousness to the lives of others. This is expressed in our social shows of respect for privacy—for instance, turning away from someone relieving themselves in public or struggling with an embarrassing situation. What comforts those to whom this show of privacy is made has nothing to do with the transmission of information—they know that we know, of course—but rather with the pantomime of obliviousness and the message it expresses. Not "I won't tell anyone about this" but "It is as if it never happened."

I am in my apartment, trying to sleep, while my neighbors are having an argument on the other side of the wall. Or maybe it's something else. We wouldn't say that I have invaded their privacy simply by overhearing them through the wall, as we would if I had planted a clandestine microphone underneath their bed. The matter is hardly exhausted by the question of physical invasion. Our evaluation changes if I do not only overhear but also shush someone who is talking, turn off the radio, put my ear to the wall. Maybe I turn on a microphone in my own room or keep a journal in which I note all that I hear from the across the wall. I have not violated my neighbors' privacy as I would have by planting the bug or drilling a hole in the wall, but we cannot say that I am really respecting it, either. The difference will not turn on the question of deliberate action—I can act deliberately in tuning them out, too, or in not keeping a record when I would really like to. Rather, I am no longer respecting my neighbors' privacy, or at least not as fully as I was before, because I have exchanged an attitude of deliberate (if imperfect) obliviousness for one of suspicion, prurient curiosity, information gathering, and documentation.

Nor is it uncommon to be on the other end of this type of experience. Perhaps you have made an acquaintance who turned out to be rather more pushily inquisitive than she had seemed at first. The typical experience of two strangers getting to know one another shifts into a different key, and you have the feeling that your interlocutor is trying to get something out of you, that she has some agenda (a hidden agenda, note, not a private one). The stranger is no longer treating you as an unknown quality whom one can get to know in the conventional, fumbling way we do when we are oblivious to the contours of another's life, but rather as someone who has something to hide. We resent and resist this shift in the relation not because we do, in fact, have anything to hide, but because we feel disrespected when someone treats us as a repository of information to be got at rather than a human being whose depths are unknown and respected as such.

There is much more to say about privacy and oblivion, and I am eager to begin the discussion in earnest. So let me give a quick overview of the structure of this book and its arguments.

Chapter 1 begins with an excavation of the origins of modern privacy at the site of its first large-scale moral panic in reaction to the rapid popular adoption of snapshot photography. This, in turn, will help to us to see and then to answer a theoretical question that has been hiding in plain sight: In what sense are invasions of privacy actually invasive? Popular moral discourse characterizes a great many things as invasions of privacy: when someone takes a picture through a bedroom window, when the government deploys facial-recognition surveillance at a public sporting event, and when insurance companies monitor our behavior to adjust their rates, to take just a few examples.[23] To answer this question, we will look back to the formative moment for our ideas about privacy and privacy rights in the second half of the nineteenth century. Our approach will be both historical and philosophical, and our way in is through a puzzle: How could the camera invade one's privacy when the naked eye did not? Understanding the logic of this invasion sheds light on the origins of modern privacy as a reaction to a technological disruption in the practices of knowledge of others and ourselves, and therefore offers insight into analogous disruptions in the twenty-first century. The originary view that a privacy invasion is possible both at home and in the street offers tools for understanding our own time, when our moral intuitions about invasiveness are confronted with the apparently public arena in which they take place, be it on social media, CCTV surveillance, or facial-recognition and GPS tracking.

Chapter 2 begins with an example drawn from Gay Talese's book *The Voyeur's Motel*, which tells the story of a man named Gerald Foos who bought a motel just so he could spy on its occupants. Foos watched his guests for decades without anyone ever finding out. When confronted, Foos insisted that he had done nothing wrong: "there's no invasion of privacy if no one complains."[24] Talese's story presents a challenge to common ideas of privacy because it seems clear that the victims of Foos's spying never suffered any material or psychological effects from his violation of their privacy. Must we then understand Foos's actions as a bit of harmless wrongdoing, from which it is but a short step to "no harm, no foul"? Similar challenges reappear with higher stakes whenever we learn of some program of state or corporate surveillance that collected information about us without our knowledge but never used it.

The common intuition that it is bad for us to have our privacy violated irrespective of what consequences ensue (even good ones!) indicates that to a significant extent we think that it is bad to have our privacy violated even if we never find out about it. This chapter, as well as chapters 4 and 5, give reasons to think that we are right about this. When Foos mounts the peephole, he does not necessarily gain any new information about his victims. He does, however, deprive them of something they would have had but for his peeping: a kind of generalized oblivion that we reasonably expect from a hotel room's four walls. One view of the interest we have in such oblivion stems from how it permits us to be unaccountable for periods of time, which is important for the accountability of agency, but also for human well-being more broadly. Privacy is valuable, we will see, because of the role it plays in the coming apart of personal identity and self-knowledge, but also because of the support it provides for meaningful forms of self-relation beyond the bounds of self-knowledge, self-control, and personal identity.

In chapter 3, we will dismantle the idea that privacy is for those who have something to hide by exploring the epistemic, normative, and existential gulf that separates hiding from privacy. In the wake of the terrorist attacks of September 11, 2001—as in many other instances of emergency and exception—the question of why insist on privacy if we have nothing to hide was one that many found hard to answer. To respond by saying "I have nothing to hide" is at once to expose oneself and to invite surveillance. Others suggested that everyone has *something* to hide, that no closet is without its skeletons.[25] This may be true, but it is a weak response that capitulates to the coercive bad faith of propaganda. Privacy, as we have already noticed, is for something other than what hiding is for. This chapter suggests a better response by using the conceptual, normative, and phenomenological differences between hiding and privacy to draw conclusions about privacy and the value of oblivion. These distinctions form part of my critique of the views of "informational privacy" by making the case that those views are, in the main, arguments for secrecy and confidentiality, not privacy.[26] As with hiding, secrecy and privacy have certain mutual exclusivities at the conceptual, normative, and phenomenological levels. From this we see that the value of privacy must have to do with more than the merely keeping oneself, an object, or information away from others, be they a single seeker or the world at large.

The comparison of privacy with hiding reveals that there are experiential or psychological dimensions to privacy as well. While the experience of hiding involves, among other things, a fixation on the world beyond the barriers of one's concealment, privacy must be characterized by the absence of this condition. The claim that privacy lacks the psychological orientation of being tethered to the world beyond one's concealment sheds light on a dark irony of the information age—that although we are more connected to one another than ever, we also feel lonelier, more socially isolated and alienated. The distinction between the experiential dimensions of hiding and privacy gives one explanation as to why. Hiding is the isolating experience par excellence, and to the extent that life in private begins to resemble hiding, we should expect that the erstwhile healthy effects of privacy would give over to the isolating and alienating effects of hiding.

After he went blind, the great writer Jorge Luis Borges said to an interviewer: "You have to remember and you have to forget. You should go in for a blending of the two elements, no? Memory and oblivion, and we call that imagination."[27] We must imagine the sort of courage required for a man whose only connection to the visual realm resides in memory—who, as he said at the time, "live[s] in memory"—to hold oblivion in equal esteem with memory, and not to think mournfully that with each vanished memory his world, his self, and his connection to them grows that much smaller and bereft. In chapter 4, we will take Borges's claim as seriously as his conviction requires, and we will do so by a route that surely would have pleased an author who so deeply and consistently troubled the line between fact and fiction. In the past several decades, scientists, philosophers, lawyers, theologians, and practically anyone else writing on the value of forgetting has taken for their touchstone a little story that Borges first published in an Argentinian newspaper in 1942. These writers almost inevitably treat Borges's story if it were a case study of the bio-cognitive limits of memory, and not a tragic fable about a boy cursed to perceive and remember everything. By virtue of becoming a commonplace, this error in interpretation offers an extremely revealing window into the value of memory and oblivion in our time.

This chapter expands the arguments of the previous ones by approaching the value of oblivion from the perspective of memory and history. We will consider the importance of forgetting for individual well-being, and then we will turn to recent debates over the right to be forgotten. The right to be forgotten is often described as a right of privacy, which is curious since its powers apply only to information that is already publicly accessible on the Internet. However, both rights share a concern with those aspects of human agency and well-being that stand in opposition to the fixity of information. Whereas the oblivion of privacy allows individuals to come apart from themselves in the present moment, the right to be forgotten allows them to sever the links connecting their present to their past selves. Finally, we will look to the perennial privacy concern about the increasing documentation of human life—what today we would call "datafication," but which has been a running concern of the public moral discourse on privacy since the value's inception. All three perspectives confirm the claims of the previous chapters: that limits to what can be known about oneself and others, and the reliable presence of oblivion in a society, play a vital role in human life, whether concerning the past, the present, or the future.

Chapter 5 brings the various strands of argument together to expand our view of the individual and public goods of oblivion. We begin with a challenge from Hannah Arendt. Across several of her published works and correspondence, Arendt describes the oblivion of privacy as a kind of social death for those confined there, but also as a necessary condition for the flourishing of individuals and the public realm itself. How can this be? This chapter engages Arendt's claim and then moves beyond it to argue that oblivion—whether produced by privacy, social forgetting, or ideas about the self—is a necessary condition for the sense that our lives, relationships, and collective projects have depth, meaning, and an inexhaustible capacity for change and renovation.

Limits to self-knowledge are responsible for the sense that there is more to our lives than meets the eye, which in turn forms the basis for the belief that as human beings we have inner resources to call upon in response to new challenges or the desire to change our lives and be different from how we are or once were. This is not a metaphysical claim about the self but one about the social cultivation of a particular idea of the person and therefore the production of individuals of one sort rather than another, with distinct capabilities and prospects for flourishing. These inner resources and depths depend upon our understanding of the human person as being at some level impenetrable to knowledge and that what lies behind that barrier, as with privacy, is not secrets or information, but oblivion.

The same holds true for the public world of collective life. There are public goods of oblivion, as of privacy. Privacy, forgetting, and other forms of oblivion's social reproduction are responsible for the sense that a human life—like a poem, film, or painting—can be called deep rather than shallow. Oblivion, as opposed to secrecy or hiding, is also essential for the cultivation of trust, both of others and oneself, and the sense that one is worthy of being trusted and therefore deserving of respect and esteem. Public life and political action also depend upon a background of oblivion for their vital qualities of meaning, possibility, and natality. Indeed, our sense of life's meaningfulness—that is, that our lives are not just ours but are also worth living—depends upon the reproduction of oblivion in our societies and selves, most notably by privacy, but also by other means. The social cultivation of oblivion contributes to the sense that we, as individuals, and the lives we live together contain unfathomable depths that cannot be called up or accessed at will, and therefore that we are at some level fundamentally resistant to instrumentalization. For what cannot be exhaustively known, recalled, or predicted cannot be entirely controlled. In these ways and more, the presence of oblivion in our lives lends our relations with ourselves and others a quality of inexhaustibility in a world that is so frequently exhausting. We must cultivate and protect spaces for oblivion in ourselves, our relationships, and our social surround if we wish to maintain the human world as a human place worth inhabiting.

1

PHOTOGRAPHY AND THE

INVENTION OF PRIVACY

We see a woman through her bedroom window. She sits with her back to us in a rocking chair. In her lap she holds an oversized teddy bear of the sort won at fairs or purchased in airports: too large to be a child's constant companion, more likely a guilty or loving extravagance from a parent who is absent all too often. Although we cannot see the woman's face, it is clear that she is well past the stage of life in which she might have ordinarily sought companionship from stuffed animals. The face of the bear, however, is visible to us: its eyes appear to seek those of the woman who, for her part, looks away from us and toward the open door at the far end of her room. From the divot where her thumb presses into the soft fur of the bear's belly, from the tensed muscles and raised veins in the back of her hand, we can see that she is squeezing it rather hard.

We see other people through other windows. We see them as they nap or assemble furniture. A leg. A woman on all fours. Faces with eyes obscured by a steel casement. The back of a woman dressed in exercise clothes smushed between the window and a Christmas tree.

We see these people in—but also through—a series of photographs taken by the artist Arene Svenson from the window of his apartment in lower Manhattan. For several years and with the aid of a telephoto lens, Svenson photographed the residents of the glass-fronted high-rise across the street without their knowledge. The neighbors might never have known these photos existed if Svenson hadn't exhibited them at a nearby gallery in 2015. The

Figure 1.1. "Neighbors #13" by Arne Svenson. Courtesy of Arne Svenson and Robert Klein Gallery.

show received attention in the press and earned Svenson a profile in *The New Yorker* magazine in part, to be fair, because the pictures themselves are quite appealing. They display a sly feeling for form and proportion; the strict geometry of the building's windows strikes a surprising, almost moving harmony with the muted unselfconsciousness of the human figures they frame. There is a sense of humor in the images that verges on the schoolboy's daring cheek. But the truth is that Svenson could have achieved these effects just as well or better had the pictures been staged. Their real appeal has to do with the candidness of their subjects (a cliché we will come to understand better in a moment) and how they invite the viewer to naughtily toe the moral line that separates an invasion of privacy from an innocent glance. In the show's catalogue, Svenson guards himself against moral reproach with some standard art-world language about capturing human universals in the quotidian, which only testifies to how obvious it is that a great deal of his images' power derives from the titillating sense of taboo and mystery that comes from peering into the hidden depths of the lives of real, particular others.

For their part, Svenson's subjects found little that was funny or profound about the pictures. Nor did they perceive any moral ambiguity about what

he had done. Even in New York City—where, as in any dense urban environment, the boundaries of the private home are fairly permeable, and it
is common to overhear one's neighbors though the walls and to see into
their windows without seeming to invade their privacy—the neighbors were
unanimous. Svenson had crossed a line. Svenson sought to exonerate himself
by analogy to the city dweller's somewhat constrained expectations of privacy: "For my subjects, there is no question of privacy. They are performing
behind a transparent scrim on a stage of their own creation with the curtain
raised high."[1] Svenson's argument is like this: to appear in an open window
when one could have closed the blinds is tantamount to appearing on the
public street; one consents to be seen by appearing on the street; consent
to be seen entails consent to be photographed; therefore those who leave
their windows uncovered tacitly consent to having their pictures taken. The
neighbors didn't buy it. "I think there's an understanding that when you live
here with glass windows, there will be straying eyes," said one. "But it feels
different with someone who has a camera."[2]

   Two of the neighbors sued Svenson to prevent exhibition of the photographs on the grounds that, among other things, he had invaded their privacy. And although on appeal the court found that the law was on Svenson's
side (on the grounds that his photographs were works of art whose display is
protected by the First Amendment), it also made clear that it thought Svenson was in the wrong and that the neighbors had suffered a moral injury and
invasion of their privacy. The court held that Svenson's photographs constituted a "technological home invasion" and lamented that although "many
people would be rightfully offended by the intrusive manner in which the
photographs were taken," the law of New York was unable to vindicate this
invasion or protect against the "heightened threats to privacy posed by new
and ever more invasive technologies."[3]

   By describing the photographs as "intrusive" and a literal "home invasion," the court, like the neighbors themselves, expressed one of our most
common ideas about privacy. It also happens to be one of privacy's more
revealing puzzles, which seems to have largely evaded attention by hiding,
like Poe's purloined letter, in plain sight. For the fact is that Svenson did
not actually invade the homes of his neighbors across the street. Like James
Stewart's character in *Rear Window*, Svenson never left his apartment but
merely captured the light streaming out of his subjects' windows and into
the world at large.

The common sense that photographing someone though their window—though also in the street and, as we will see, even as they are performing on stage—is *intrusive* or *invasive* is both natural and deeply weird. This language persists because it expresses something of how it feels to be seen under such circumstances. If I look out my window and see someone looking in (note the prepositions) or taking a photograph, it really does feel like an invasion: the barrier between us disappears in an instant and the Peeping Tom or photographer is inside my home, my private space. And not just inside, but close enough to touch. A boundary has been crossed that is not merely a moral boundary, but also a psycho-spatial one having to do with the delimited zone of privacy. The sense that a gaze or a camera lens goes into the private space also reflects a lived experience of vision in which our sight travels out into the world. This is the natural or naive way we experience sight, even though to-day we know perfectly well that everything we see arrives to our eyes as rays of light.[4] We look into matters, boxes, eyes. We watch the action *closely from a distance*. We see through binoculars, glasses, trees, lies. Stares pierce, eyes probe, and gazes can be felt upon on the skin. "The pressure of their stares on my back . . ." "His eyes were all over me." "Eyes on your own paper!"

If we wish to understand this way of thinking and its connection to priva-cy, we should first pause to appreciate just how strange it is. For, pace the New York State Court of Appeals, Svenson never actually entered into his neigh-bors' apartments. As a physical matter, he neither intrudes nor invades, but merely captures the light flowing out through an uncovered window. If he were *actually* to invade the homes of his neighbors, we'd call that trespassing or breaking and entering, not a privacy violation (burglars, for instance, are never charged with invading their victims' privacy in addition to breaking their locks and stealing their things). Moreover, it is not at all obvious why Svenson's photographing his neighbors through their windows should con-stitute an invasion when, according to the mores of urban living and even the neighbors themselves, simply looking in would not. The difference cannot be that Svenson's photographs revealed any secrets or that he used an unusually powerful tool to capture them, something more like an X-ray than a snapshot camera. What he saw through his lens was not only anodyne but also visible with the naked eye or a pair of binoculars from his living room or the busy street below. Anyway, the neighbors' complaint has nothing to do with the type of equipment he used; their objections wouldn't vanish if he had taken the same pictures from just outside their windows with a less powerful lens.

The puzzle of the eye that invades, as I have begun to sketch it, is more than a mere trick of language or legacy of dead metaphor. It goes to the very heart of the modern value of privacy. To see how, we will turn to the period in which the modern conception of privacy first took shape, in the second half of the nineteenth century, when the invention and rapid spread of photography changed the way human beings came to know one another, themselves, and the world. It was then that the modern sense of privacy's perceptual invasion was born in earnest; the ideas and concerns of that era are still ours to a large extent. When the Svenson court called upon the New York state legislature to revise the privacy laws to respond to "these times of heightened threats to privacy posed by new and ever more invasive technologies," they were invoking a trope that is at once perennial and original to the modern idea of privacy. For instance, the foundational text of the right to privacy, Samuel Warren and Louis Brandeis's 1890 "The Right to Privacy," said basically the same thing about the same technology: "Instantaneous photographs and newspaper enterprise have invaded the sacred precincts of private and domestic life; and numerous mechanical devices threaten to make good the prediction that 'what is whispered in the closet shall be proclaimed from the house-tops.'"[5] Although the common sense about privacy's value may have shifted in the century that separates us from its earliest advocates, the understanding of the threats it faces have been strikingly consistent. Thus, the history of early privacy advocacy offers an estranging yet deeply revealing mirror in which we may see our own ideas about the value and challenges of privacy with fresh eyes.

Modern privacy's point of origin is not just continuous with contemporary concerns about concealment and exposure. Seen from the present moment, it also appears as a site of missed opportunity. Those who lived through the introduction of photography into social life argued for, among other things, a value of privacy that did not expire the moment one walked into public or opened the drapes. As we will see, they advocated for a view of privacy concerned with protecting realms of unarticulated potentiality in both individuals and society at large, and not in the first instance with the importance of keeping secrets or maintaining a respectable public face. This view is not only relevant to and clarifying of today's privacy concerns, but by not ignoring the moral and social question about the sheer existence of information in human affairs, it presents us a more powerful alternative for defending private life in the era of big data, algorithmic influence, and the

"Internet of things." Fortunately, this view of privacy is not lost completely; it still animates much of our moral thinking about privacy and undergirds a good deal of meaning in human life. It is still there, lurking in the puzzle of the photographic invasion of privacy.

The Modern Demand

The Victorian era was the first to witness large-scale social anxiety about privacy. Their moral panic resembles the many that succeeded it—including our own—in that they, too, were convinced that privacy was under new and unprecedented threat in their time. Unlike us, however, they also thought that privacy itself was new, a quintessentially modern reaction to upheavals in their social world. Writers on the subject were consistently obliged to note that "privacy is a distinctly modern product" and that a right to protect it was one of the "new demands of society."[6] This widespread insistence on privacy's novelty is striking in light of the longstanding and cross-cultural existence of norms surrounding concealment and exposure. To be sure, the earliest advocates for the importance of privacy and privacy rights did not suppose that theirs was the first generation in history to place moral weight on concealing certain aspects of human life. Instead, their insistence on privacy's novelty was meant to assert a break with the normative ideas about concealment they had inherited and to distinguish the modern interest in privacy from older related interests, like those in property, secrets, confidence, and anonymity.

It is easy to overlook the novelty of the modern view, notwithstanding the constant assertions of its early advocates. The concept of *the private*, meaning that which does not pertain to the state or common good, has been common currency since the Roman res publica, whose legal distinction between what was *publicus* (pertaining to the state, collective, and common good) and *privatus* (pertaining only to the individual) was given second life in the modern era with the rise of liberal political philosophy and political economy. Likewise, English usage of the adjacent idiom of "privy" to refer to hidden, secret, or isolated places (most commonly a toilet) and analogous sorts of knowledge, like secrets (to which one can be privy), dates more or less to the earliest days of the language.[7] And although it is important to mark the deliberate use of the relatively new noun "privacy" to distinguish the concept from the older nominals "private" and "privy," normative use of "privacy" in English dates back at least as far as the 1740s. Indeed, before the middle of the nineteenth century, privacy was used to refer to several interests in the vicinity of the modern idea, such as the nondisclosure of unpublished writings and the physical boundaries of real property. For instance, in 1814 an English court held that although a "defendant might not object to a small window looking into his yard, a larger one might be very inconvenient to him, by disturbing his privacy, and enabling people to come through his property."[8] The issue here was that someone could physically trespass onto the property, and not merely look in; otherwise the defendant would have the same reason to object to a smaller window as to a larger one. (At the time "window" could refer to any sort of opening in a wall that admitted light and air, including what we now call doorways, which explains how Hawthorne could open one of the more heartrending scenes in *The Scarlet Letter* with the sentence "Governor Bellingham stepped through the window into the hall, followed by his three guests" without intending the comedic effect that the line carries today.[9]) There was nothing particularly novel about this kind of privacy because it is essentially synonymous with the protections of private property. There was no need to argue over what privacy was or to advocate for a special right to it, for a violation or invasion of privacy in the eighteenth and early nineteenth centuries was already a violation or invasion of some other right or interest.[10]

A new and distinctly modern value of privacy begins to show up around the middle of the nineteenth century. An 1850 newspaper column in *The Independent* titled "The Sacred Privacy of Home" is emblematic of the new conception:

> It is the calamity of the poor in great cities that they cannot enjoy the seclusion of a home, but must occupy a mere place in a crowded tenement. . . . This promiscuous herding of men, women and children is contrary to nature and is unfavorable to social and moral cultivation.[11]

The sacredness of this privacy is not a function of the bourgeois value of the domestic home and its associated interests in property, nor of the instrumental value of keeping secrets or maintaining confidence, but a condition of human well-being as such. The tenement's lack of privacy is harmful because it interferes with personal development (even in adults) in a way that is detrimental to a fundamental interest. To be sure, the notion that the misfortune of the newly urbanized prole was that he had roommates expresses the suddenly elevated importance of privacy to the nineteenth-century bourgeoisie as much as it belies the moral myopia of the patrician class. Yet this sacred privacy and cultivation was not of value for the cultivated classes alone, as might be the pleasures of solitude or tranquil reflection, the sort of thing that requires a sophisticated character or a country estate. Notwithstanding that most privacy advocates in the archive are bourgeois, if only because theirs were the concerns that made it into print, the value of privacy for which they argue is not that of the walls of the private home that keep the urban moil at bay. Rather, privacy was thought to be something in which everyone had an interest solely by virtue of being human.

This universal human interest in privacy emerged against a background of rapid urbanization in the nineteenth century, the Enlightenment's enduring legacy of human universals and emphasis on individual rights, and above all the development of modern notions of the self out of Romantic ideas about individuality and personality's moral worth.[12] This view of the self—what it means to have one, what makes one distinct from another, both empirically and morally, and so on—is so common today, albeit detached from Romantic metaphysics, that it hardly bears description. The basic idea finds expression in the concept of individual personality, which is premised upon the notion that part of what it means to be a human person is to have an individual way of being. If the kind of person one turned out to be was up to oneself rather than being imposed by nature, then people could no longer be divided into natural types but had to be perceived and interpreted in their appearances, words, and deeds to make out what sort of person they were—by others but also by themselves. This view of personality was individual in two important senses: it differed from person to person and it was up to the individual to fashion it, either as an expression of one's authentic inner nature or, as would come somewhat later, deliberately according to one's will.[13] Such elevated freedom of individual self-fashioning also entailed new responsibilities and obligations. The possibility of freely developing oneself out of one's potential implies the necessity of doing so, now that individuals were thought to be responsible for who they turned out to be, both in ethical terms and in the broader, aesthetic dimension of "personality."

The idea that others, from strangers to intimates, were more like characters in a novel than stock figures or natural types introduced a powerful social incentive to scrutinize the behavior, appearance, and closely associated objects of others. This incentive appears throughout our aesthetic and ethical vocabulary for thinking about the self, much of which has its roots in precisely this era: for instance, the "stereotype" as a kind of ready-made personality (a metaphor drawn from contemporary innovations in newspaper printing) and the "interview" as a kind of questioning that arrives at the truth about an individual not by examining the historical record but by *seeing into* someone and *eliciting* revelations from within them. (This view of personality is what motivates the common judgment that the best interviewers are those able to draw out information that was either unknown or unacknowledged by their subjects.) The burgeoning urban centers of the nineteenth century provided extremely fertile ground for this detective work of the self, and before long it became a quotidian experience and expectation of social life.[14] These new practices of perception and knowledge came with newly heightened anxieties surrounding perception of persons and their effects, which in turn contributed to the homogenization of public appearances in the Victorian era.[15] This backlash, however, only reinforced the idea that accurate, authentic knowledge of a person is achieved by *seeing through* the protective surface of convention and *into* the deeper truth concealed within.

The new imperative to scrutinize others in order to discover what they are really like is mirrored in the domain of self-knowledge as an imperative to distinguish one's real or authentic personality from the conforming pressures of society. The Romantic and post-Romantic answer to this challenge was to associate nonconformity with autonomy and to trust above all else in the spontaneous expressions of thought and behavior as reflections of what one was really like or thought. The column on the sacred privacy of the home continues:

> Honored and cherished be the privacy of home; there let the man become a boy again, and the dignified statesman and the grave divine without scandal participate in the sports of childhood, down upon all-fours at a game of marbles, or of cost for a game of ball; let the notes of love and glee ring out as nature prompts them, without affectation and without prudishness.[16]

The value of the statesman and divine's privacy is not primarily a function of its protection against scandal, although that might be a welcome byproduct. Instead, it lies in marking off a space and time in which they can act according to the spontaneous dictates of their inner nature, which fear of scandal might constrain or stifle. The statesman and divine cannot know ahead of time how nature will prompt them, and therefore they cannot say in advance what they will come to know about themselves. It might be something they'd rather others not know; in private they can recognize and face the spontaneous emanations of personality and then decide whether or not to keep them secret. Of course, what they discover about themselves might turn out to be utterly conventional and unsurprising. There's just no way to know for sure in advance what one will spontaneously express. This is because personality's core, on this view, consists of unformed, indefinite potentiality.

Ralph Waldo Emerson made the same point in his famous essay on self-reliance, published nine years earlier. Serious inquiry into the essence of human personality—into, as Emerson put it, the "deep force" of "the aboriginal Self. . . . that science-baffling star, without parallax, *without calculable elements*"—"leads us to that source, at once the essence of genius, of virtue, and of life, which we call Spontaneity."[17] What the statesman and the divine's privacy protects and indeed produces is, among whatever else, a zone of potentiality at once correlative and conducive to the spontaneous expression of personality's potential. The home's reliable barriers to perception are valuable on this view—"sacred," even—because of the confidence they give the statesman and divine that they can act as nature moves them without having to worry about being made to answer for every emanation.[18] They can experience the free play of personality's inner potential undampened by fear of having to account for spontaneous expressions at odds with their typical comportment or public persona.

If spontaneous emanations from the self's interior were thought to convey the most authentic truths about an individual, the face was the best and most reliable place to read them. Belief in the science, now pseudoscience, of physiognomy, whose central idea was that physical appearance, and especially facial expression, offered a window to the truth about personality, was widespread at the time. Guillaume Duchenne's use of electroshock to produce the array of common facial expressions was thought by many, including Charles Darwin, to have uncovered a mechanistic grammar of expression in the involuntary movements of the face.[19] Whereas the "guarded" face was a mask capable of change at the wearer's direction, the unguarded face was a sort of camera in reverse: a direct, mechanistic expression of interior states. "A man moderately angry, or even when enraged, may command the movements of his body," writes Darwin in one of the first books to present photographs as scientific evidence, "but . . . those muscles of the face which are least obedient to the will, will sometimes alone betray a slight and passing emotion."[20] Arthur Schopenhauer makes the point more clearly: "The face of a man gives us fuller and more interesting information than his tongue; for his face is the compendium of all he will ever say, as it is the one record of all his thoughts and endeavors . . . the outer man is a picture of the inner, and the face is an expression and revelation of the whole character."[21] This way of thinking persists in the emotional and epistemic value we still place on unposed photographs as being more real ("Look natural!"), expressive, or authentic, and the use of so-called body language experts to explain what a politician really meant in his speech.

The expressiveness of personal appearances opens a route for truths about individuals to pass from the self's interior into the realm of public perception independently of an individual's will. I blush in response to a question about my love life, and I seem to have betrayed myself by revealing something I did not wish to disclose. I might also have learned something about myself. It is significant that we call the sorts of questions that elicit this reaction *invasive questions*, to which the standard response is "that's private." We will have more to say about the invasive question at the end of this chapter, although here we should note that questions of a personal nature surely predate the nineteenth century's moral panic about privacy invasion. Something changed, however, with the invention of the photograph camera.

The Speaking Likeness

Toward the end of the nineteenth century, the slow buildup of modern ideas about privacy exploded into moral panic. Headlines like "Is There Any Privacy?," "No More Privacy," "The Decay of Privacy," and "Is the End of Privacy Coming to Human Kind?" began to appear with increasing frequency on both sides of the Atlantic.[22] These headlines have a familiar ring to them. It is hard not to recognize a quality of our own historical moment in their hyperbolic tone of concern and eulogy, even if they are not from the turn of our century but the previous one (1874–1928). Their advent in the popular press marked the beginning of a period that stretches into the present day in which the death of privacy is constantly proclaimed, often in apocalyptic terms. This is another way that modern privacy broke with previous usages: it was the first form of privacy that is always dying. However strange the Victorians may appear to us, we share with them this sense of privacy as something constantly on the brink of extinction.

Victorian privacy advocates were animated by what they saw as a radical shift in the epistemic practices by which people came to know one another and themselves. The general assertion that "new relations give rise to or develop new or inchoate rights" was a commonplace in their writings, which were consistently clear about what they saw as the principal causes of those new relations: the development of a culture of publicity enabled by the rise of the first mass media in the form of the newspaper industry, and the invention of the snapshot camera.[23] The immense success of Warren and Brandeis's article is a testament to how well it expressed and consolidated the common thinking on this score:

> Instantaneous photographs and newspaper enterprise have invaded the sacred precincts of private and domestic life; and numerous mechanical devices threaten to make good the prediction that "what is whispered in the closet shall be proclaimed from the housetops."[24]

To begin to see the strangeness of this formulation, we should note that the new threat of invasion is not, in the first place, what is today typically understood as the main danger associated with privacy. Their concern was not that the invention of photography brought about a new threat of breaching one's concealment and gaining some bit of information which could then communicated to the wider world. There was nothing novel about the threat of words spoken in confidence or behind closed doors making their way to ears beyond their intended audience. The line about the closet and housetops comes from the Bible, and the danger of which it warns is as old as human society. By itself, this would hardly have warranted the recognition of new moral and legal rights, not least because nineteenth-century privacy advocacy took place against a background of longstanding norms and laws against eavesdropping, gossiping, confidence breaking, and the revelation of secrets.[25] Nor was the camera's novel invasiveness a function of some new power to reveal what is hidden, to peer into the secrecy of the whispering closet. To be sure, telephoto lenses like Arne Svenson's and infrared or X-ray film may permit us to see what was otherwise invisible to the naked eye. Yet Warren and Brandeis and their contemporaries were not concerned with these potentially invasive aspects of the technology, in part because they were either exceedingly rare or yet to be invented, and in part because by itself a powerful lens is no different from a telescope or pair of binoculars, whose use long predates demands for a right to privacy but which are never mentioned as relevant threats.

The complaint about "instantaneous photographs" responds specifically to the development of the handheld snapshot cameras that, by today's standards, were still fairly primitive. To get a good picture, one still needed to be within ordinary and unaided eyeshot. There was also widespread concern about the invasiveness of tiny (for the time) "detective cameras" that could be disguised as vest buttons, hats, packages, and, in a rather cheeky, evocative instance, a pistol.[26] But this only exacerbates the puzzle of invasion, since these cameras could only photograph what their wearers had already seen with the naked eye, typically *in public*; they were designed to capture not the hidden but the unguarded and spontaneous. To understand how cameras were thought to present a novel threat of invasion when binoculars or eavesdroppers did not, we will need to look closely at the features of the new technology that were salient to nineteenth-century privacy advocates and understand how those features were thought, as Warren and Brandeis put it, to subject individuals "to mental pain and distress, far greater than could be inflicted by mere bodily injury."[27]

The sudden introduction and rapid spread of photography throughout the industrialized world was a watershed in the history of the knowledge of self and other on the order of the printing press and the Internet. It is hard to appreciate just how radical the advent of photography was for those who lived through it. Edgar Allan Poe expressed a common sentiment when he called the photograph "the most important, and perhaps most extraordinary triumph of modern science."[28] Oliver Wendell Holmes thought it "the greatest of all human triumphs over earthly conditions."[29] What excited and astounded was above all "the camera's innate honesty," its seemingly objective faculty of representation.[30] For the first time in human history it seemed possible to take a piece of the world and preserve it, just as it was, unmediated by human interpretation, bias, or the many limitations of sensory perception and recollection.[31] There was a sense of the Promethean about this seeming usurpation of divine power over time and appearance, even death, by sheer human cleverness. The result was a revolution for the relationship between vision and knowledge:

> All language must fall short of conveying any just idea of the truth. . . .
> Perhaps if we imagine the distinctness with which an object is

reflected in a positively perfect mirror, we come as near the reality as by any other means. . . . If we examine a work of ordinary art, by means of a powerful microscope, all traces of resemblance to nature will disappear—but the closest scrutiny of the photogenic drawing discloses only a more absolute truth, a more perfect identity of aspect with the thing represented.[32]

Poe reveals a truth about how we see and value photographs by telling a lie about the medium: if you look closely enough at a photographic print you will not see the object pictured there more clearly but grains of silver-ions in emulsion or, today, pixels. Nor were early viewers of photographs so naive as Poe's enthusiasm might lead us to believe: at first they found the photograph's flattened perspective decidedly unnatural, and it was common knowledge (indeed, a popular demand) that prints and negatives were frequently altered.[33] Nevertheless, buildings were said to have drawn themselves with "the pencil of nature" in "the mirror with a memory."[34]

We like to think of ourselves as savvier than photography's early rhapsodists, yet we still unthinkingly treat the photograph as if it were a window through time and space, through which we literally see Abraham Lincoln, our dead relatives, ourselves as children. "Look, there I am! That's me at five." If I show you a photograph of sand grains under heavy magnification that reveals them to be completely unlike the sand you know from experience—more like little galaxies than granules (Fig. 1.2)—you will believe me that this is what sand is *really like* not because you trust me, but because you trust photography. This has less to do with the medium itself than with the epistemic presumptions we have developed around it, how we trust photographs to show us the world in the same way that we trust eyeglasses and binoculars but not paintings and verbal testimony. If this seems old fashioned in an age of digital manipulation, consider our unwavering faith in the veracity of X-ray photos, as marvelous at the turn of the last century as they are commonplace today. Though you might want to have your bones photographed in a newer or more powerful machine, it would be absurd to go to another machine for a second opinion. And although it might seem like the advent of "deep fake" images produced by artificial intelligence spells the end of this epistemic faith in the photograph, the opposite is true. The very idea of a "deep fake" image depends upon and further solidifies the counter-category of the "deep real" picture.

Figure 1.2. "Sand grains, Maui," by Gary Greenberg. A magnified photograph of sand grains reveals a marvelous world invisible to the naked eye. Courtesy of Gary Greenberg, www.sandgrains.com.

Given the technology's rapid proliferation and commercial success, the new epistemic faith in photographs had a seismic effect on the relationship between vision and knowledge in daily life. "We see everything and everybody in the maze of distracting cross-lights," wrote one privacy advocate.[35] The history of human subjectivity is of course marked by such transformations in habits of perception and the social and political consequences that ensue. Consider, for example, the invention of ideas of race and how its epidermal visibility subsequently structured the intersubjective life of racist societies—as good a reminder as any that perceptual transformations will typically have moral and political ramifications, too.

Paul Valéry recalls how the shift in modes of perception brought about by photography fundamentally altered the terrain of moral and social life:

> A marked revision occurred in all standards of visual sight. Man's way of seeing began to change, and even his way of living felt the repercussions of this novelty, which immediately passed from the laboratory into everyday use, creating new needs and hitherto un-imagined customs.[36]

A revision in the received habits and standards of sight is an epistemic rupture because it entails an alteration of the status quo relationship of vision to truth, be it knowledge of history, the natural world, others, or oneself. Valéry elaborates: "Prior to [photography's] invention, any fact, provided a sufficient number of people swore that they had seen it with their own eyes, was considered incontestable." But now a single snapshot "was proof enough to demolish the testimony of some hundred people."[37] Trust in one's own experience—"I saw it with my own eyes"—was overtaken by mechanical means, and one's authority to speak for oneself about what one saw firsthand was irreversibly undermined.[38]

Prominent among Valéry's new needs and unimagined customs was the modern value of privacy. And it is in light of his epistemic revision of sight that we can understand one aspect of what was new about Warren and Brandeis's fear that "what is whispered in the closet shall be proclaimed from the house-tops." The apparent objectivity of the photograph—or the audio recording, whose development was roughly contemporaneous with that of photography, albeit with less immediate commercial success—seemed to mean that the one who is recorded is made to appear or to speak wherever the image or recording is reproduced, including situations and contexts of which the subject is unaware. This is concerning irrespective of whether what is said is secret or common knowledge, since we tend to think it important that individuals have the ability to control where and when they will appear and speak. This moral idea is the basis of Warren and Brandeis's assertion that the closest analogy to the right to privacy is not an individual's right to protect property or to control personal information, but that "of *determining*, ordinarily, to what extents his thoughts, sentiments, and emotions shall be communicated to others."[39] In other words, the harm of a privacy invasion is less like having one's secrets stolen than being forced to divulge them. What makes such an act bad—that is, both wrong and harmful—is not primarily a function of the information in question but how it undermines or violates an individual's agency.

Now we can begin to understand how it could be invasive to photograph somebody through an open window when merely looking into the window would not be. For one thing, a different form of knowledge is produced in either case. Whereas the living memory of an eye-witness is fluid, resistant to being called up at will in the same form every time, and subject to constant refashioning and the attrition of forgetting, a photograph is fixed, stable for all to see and scrutinize, more like a piece of information than a memory. This corresponds to another difference having to do with the communication of what is seen. If one looks into a neighbor's window and gives an account of what one saw, one speaks for oneself and gives one's personal point of view on what was there. But when one takes a photograph or audio recording and then shows that picture or plays the recording, it is the one pictured or recorded who appears or speaks—who is *made* to appear or speak. Warren and Brandeis were not alone in thinking that invasion of privacy was more like a forced confession than breaking and entering. The fear that the invention of photography heralded a new era of technologically forced revelations appeared with increasing frequency around the turn of the century in the form of fearful prognostications. For instance, in 1896, a writer in the *Atchison Daily Globe* worried that soon

it will be as easy to photograph the doings of our neighbors in the next room as to look out the window. Spying through keyholes will be done away. . . . There will be no sure privacy for anybody anywhere. The desperate resort of swallowing a diamond or a valuable document to conceal it will no longer avail, for the inside of the stomach can be photographed as easily as the face.

If in addition to this the invention which will photograph thought succeeds there will be nothing left for any of us except to just behave ourselves in the strictest manner. We shall not dare even to think anything wrong, for fear it will be brought up as a witness against us.

In view of the approach of this dread time, it will be a good plan now to begin to practice for it—in our minds.[40]

Or this item titled—what else?—"No More Privacy," syndicated across the United States in 1889:

> And now a meddlesome, interfering doctor has gone and invented some kind of a "scope" or something through which he looks into a fellow's eye, without asking him a solitary question, and knows right away whether or not the fellow smokes, how many cigars a day, and about the kind of cigars. It's getting to be so that it's no earthly use to lie to a doctor. And by-and-by the preachers will begin to find us out the same way.[41]

What is wrong about the forced confession or the thought-reading scope is not primarily that they reveal a secret. It is the force that makes it wrong. And not just wrong, we tend to think, but harmful. It does not make much difference to our moral judgment of the forced confession if what you are forced to confess is something that could have been discovered by asking around, or if it is something your inquisitor already knew, or a fact completely anodyne, a matter of indifference like what you ate for breakfast today. It might be worse for you if what you are forced to confess is embarrassing or can injure your prospects, but these consequences are independent of the badness of the forced confession itself; it isn't exculpatory for the inquisitor to elicit only common knowledge or flattering secrets.

To understand what the doctor's scope invades, think about the faculties of judgment and intention by which one decides what one will communicate, when, and to whom. These faculties are internal to agency in the sense of being necessary for it, but they also tend to take place in the epistemically privileged realms of self-consciousness. Moreover, there is an additional sense of internality here associated with the moral right that individuals have to make up their own minds, which corresponds to the sense in which we tend to characterize reasons that an individual does not endorse as being "external" to her. It is in just this sense that the normative gaze of judgment is said to *penetrate*: one *internalizes* the watchful eyes of a supervisor, a parent, or the disembodied eye of public opinion, by taking them into one's processes of judgment and intention. On this view, the photographic invasion of privacy does not consist in obtaining some piece of information from the human interior and bringing it outside into the realm of publicity—again, that is what the revelation of secrets does. Rather, it consists in one's morally privileged realm of agency being invaded by someone else, who then gets to decide what one reveals about oneself, what one says, where one appears, and therefore how one lives one's life and what will be true about it. There is no spatial interior that the doctor's scope *sees into* when he sees into a patient's mind (as opposed to, say, his skull or brain), no physical or optical boundary that is crossed. Rather it is a moral limit that is crossed having to do with the patient's agency and self-knowledge.

Strange as it may sound, this was the complaint of two of the very earliest attempts to enforce a legal right to privacy. The first was a civil suit brought by an actress named Marion Manola, one year before the publication of Warren and Brandeis's article. Manola sued a photographer named Benjamin Stevens who, in the middle of one of her shows, snapped a picture of Manola as she performed on stage. In interviews, Manola insisted that she was not embarrassed to be seen in costume, nor did she think that her concealment or secrecy had been broached. It was a public theater, after all; Stevens was a paying member of the audience, wholly permitted and expected to take her in with his own eyes. She even posed for publicity photos in the very same get-up that she was wearing when Stevens snapped the photo.

Rather, Manola's complaint was that by printing and displaying the photograph, the photographer forced her to appear in places and circumstances against her will. She didn't ask the court for any sort of monetary compensation, but only to have her control over where she appeared in public restored.[42] Warren and Brandeis clearly had this case in mind when describing the new form of photographic invasion: they refer to it just a few lines after the threat that "what is whispered in the closet will be shouted from the housetops" along with three separate instances of its coverage in the *New York Times*.[43] It seems that it was in part to vindicate Manola's right to privacy that they published their article.

A few years later in a similar case involving the unauthorized reproduction of a photograph, the Supreme Court of the State of Georgia would be the first in the United States to recognize a legal right to privacy. Paolo Pavesich sued the New England Mutual Life Insurance Company for reproducing a photographic portrait of him in an advertisement (Pavesich is on the left).

The negative from which the photograph was reprinted was not stolen from Pavesich—it was the property of Atlanta photographer J. Q. Adams— nor did it identify him by name or depict him in an embarrassing or shameful light. He's "the man who did," after all. Nor does the photograph reveal

Figure 1.3. A "cabinet card" photograph of Marion Manola posing (n.b.) in costume. Billy Rose Theatre Division, The New York Public Library for the Performing Arts.

anything secret or concealed. In his pose, Pavesich deliberately invokes the visual language of portraiture, by which we can assume that he expected his photograph to be seen and interpreted in a certain way. The slouch, the sidelong gaze, the rakish twisting of the mustache, and the flamboyant bow tie all seem intended to convey a sense of a thoughtful character, financially successful if perhaps a bit more eccentric and creative than your average bourgeois. (Pavesich was, it seems, something of an artist.) Nevertheless, the court concluded unanimously that the unauthorized reproduction of his photograph in a newspaper was a violation of his moral right to privacy, not

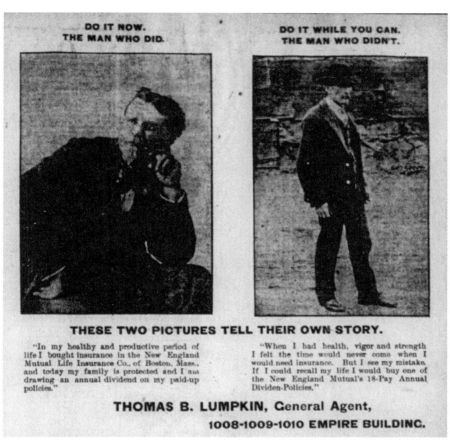

Figure 1.4. Life insurance advertisement from the Atlanta Constitution. Paolo Pavesich (left) appears in an advertisement for life insurance, printed in the November 15, 1903, edition of the Atlanta Constitution (Atlanta History Center).

because of any property-type right he had to his own image, but because it was tantamount, morally and literally, to forcing him to appear and express himself in person.

> The knowledge that one's features and form are being used for such a purpose and displayed in such places as such advertisements are often liable to be found brings not only the person of an extremely sensitive nature, but even the individual of ordinary sensibility, to a realization that his liberty has been taken away from him, and, as long as

the advertiser uses him for these purposes, he can not be otherwise than conscious of the fact that he is, for the time being, under the control of another, that he is no longer free, and that he is in reality a slave without hope of freedom, held to service by a merciless master; and if a man of true instincts, or even of ordinary sensibilities, no one can be more conscious of his complete enthrallment than he is.[44]

We are familiar with other sorts of cases in which individuals appear or speak without having intended to—like one who sleepwalks, or blurts out secrets, or who suffers the verbal tics of Tourette's syndrome. We have no problem calling these cases failures of agency. The difference between these and the photographic cases, however, is that with the photograph it is someone other than the agent who makes her appear, or speak, when she had not intended to. The proper analogy here is with the one whose clothes are torn off in public or is made to confess. We most naturally call such acts *violations*—the other term that will come to describe invasions of privacy.

I suspect the notion that the camera forces one to appear or confess in places that one's physical body is not may strike some readers as old-fashioned, a bit of superstition from the age of phrenology and spiritualism. Yet it was and remains a common trope of privacy discourse from the nineteenth century to today. As we shall see throughout this book, concerns about technological threats to privacy and their nature have been more consistent than we commonly assume. For instance, consider the moral panic over the development of databases in the second half of the twentieth century. The entry of more and more aspects of human life into documentary record was thought to pose a threat to human well-being because it contributed to "the constitution of an additional self, one that may be acted upon to the detriment of the 'real' self *without that 'real' self ever being aware of what is happening*."[45] Later, with the digital revolution, information collected about individuals was thought to constitute an actual doppelgänger called a "data double," now a term of art in surveillance studies, which again meant nothing less than "the formation and coalescence of a new type of body, a form of becoming which transcends human corporality and reduces flesh to pure information . . . a new type of individual, one composed of pure information."[46] It is hard not to hear in these statements an echo of Holmes and Poe's awe before the daguerreotype, or the Pavesich court's moral horror at the photographic doubling of the person. Like the photograph, an individual's data double is thought to be objective and forthright precisely because it speaks for its subject but in a way that bypasses the will of the living agent. Like a photograph, the data double is passive, transparent, and open to scrutiny. Like the photograph, the data double is thought to be candid or *revealing* in a way that its corresponding person may not be. Sociologists have documented how employers will trust what an applicant's data double reveals about them over the applicant's own testimony because they "tend to view applicants' data-doubles as keys to who they really are."[47] Unlike the living agent, the data double is thought to have nothing to hide, not because it is necessarily forthcoming—there always remains the ambiguous and open-ended task of interpretation—but because it is "composed of pure information."

Inviolate Personality, Stolen Potential

The moral panic over privacy in the 1880s and 1890s coincided with a mania for candid snapshots of strangers. The first mass-produced snapshot camera was George Eastman's Kodak in 1888, yet already in 1884 the *New York Times* was printing stories on the "Camera Epidemic."[48] Photographers were commonly referred to as a scourge, lunatics, and devils: "kodak fiends driving the world mad."[49] The enthusiasm for and fear of candid snapshots drew upon and perpetuated the view of unwilled bodily expression as a privileged window into the truth about an individual. Take, for instance, this encomium to the concealable camera, published in a camera-hobbyist magazine in 1888: "The beauty of the invention is that the victim is thoroughly unsuspicious. He does not know that his picture is being 'took,' consequently the character [i.e., personality] is all preserved."[50] Recall Schopenhauer's claim that "the face is an expression and revelation of the whole character." No surprise, then, to find him echoing the hidden-camera enthusiast in his essay on physiognomy, writing that "photography . . . offers the most complete satisfaction of our curiosity" about what a person is really like, far exceeding what that person might have to say about himself and providing "fuller and more interesting information than his tongue."[51]

Courts even considered photographing the accused at the moment their charge was delivered and introducing the image at trial as evidence of "the lineaments of guilt or innocence" betrayed by the face.[52] The interior of "interior states" could now be accessed without the consent of the person whose interior it was, thanks to late nineteenth-century advances in photographic technology that permitted a candid expression to be reliably captured, perhaps on the sly, separated from the stream of other expressions, and scrutinized for meaning. On this view, the camera really does seem to present a novel form of invasion, which would explain why Warren and Brandeis would argue for a "general right to privacy for thoughts, emotions, and sensations . . . whether expressed in writing, or in conduct, in conversation, in attitudes, or in *facial expression*."[53]

When Warren and Brandeis claimed that what the camera invades—that is, what the right to privacy protects—is neither secrecy, seclusion, nor property, but "inviolate personality," they are drawing on that tradition of thought about the self we discussed a moment ago, which traces back to the Romantic idea that the essence of the human person was a quality of potentiality. As Friedrich Schiller, a lifelong favorite of Brandeis, put it, "personality, considered in itself and independently of any sense-material, is merely the disposition for potentially infinite expression."[54] Closer to Warren and Brandeis's day, Emerson said more or less the same thing, more or less all the time: "In all my lectures, I have taught one doctrine, namely, the infinitude of private man."[55] The value of the divine and the statesman getting in touch with their inner children, and the danger of having one's candid image snapped in a photograph, depend equally on this idea that at the heart of the human person was a quality of unfixed potentiality, which was not just the basis for the personality one formed out of it, but an inner resource that was valuable in itself.[56]

Here we have a more profound sense in which one's "thoughts, sentiments, and emotions" may be "determined" by the photograph and the forced confession: they are made determinate. Our thoughts, sentiments, and emotions are not like secret files locked away in a safe, where the doctor's scope or a candid photograph can then discover them. Of course we all have secrets, but most of what we feel and think we discover like the doctor would with his scope: it appears to us, often suddenly, out of the blue, as it were—or, as I will argue later in this book, out of the oblivion within oneself, beyond the frontiers of self-knowledge. When we say that a feeling, thought, or expression *comes to us*, we do not mean that it emerges from some inner hiding place where it has been biding its time, fully formed, waiting for just this moment. Rather, what we mean is that it comes into existence at the moment of *and through the action of* our becoming aware of it. This is why the concept of expression (both public and private) plays such a fundamental role in

Warren and Brandeis's article, and why they are so drawn to the analogy of life in private with an artist's drafts, for which privacy is valuable in virtue of the protection it offers *against* the fixity of public expression and not, in the first place, because of how it empowers artists to exercise control over who sees them. One can discover what one thinks or feels by expressing oneself, as they put it, "by word, or by signs, in painting, by sculpture, or in music" in the same way that one expresses oneself in facial expressions and attitudes.[57] However, once these expressions attain a quality of publicity, the still-indeterminate nature of the draft is replaced with the fixity of the published work. As Nancy Rosenblum describes the Romantic roots of this view of personality: "Something is lost, not exhibited or gained, when infinite possibility gives way to the limitations of actuality."[58] This is what the doctor's scope and the camera seemed to do, except that with the scope and the camera it was someone other than the individuals at whom they were aimed who produced and fixed the truth about them.

If the domain of privacy is delimited by the distinction between fixed expression and unexpressed potentiality, and not in terms of secrecy or control (for you cannot control what you cannot know), then we have a sense of what the camera actually invades: the protean realm of personality's potential.[59] By capturing a momentary expression and preserving it to be scrutinized for meaning, the photograph crosses the threshold of personality whose division between inner and outer is of a particular epistemic sort—not a physical barrier like a house or a body, but rather one corresponding to the division between fixity and potentiality. The doctor's truth-producing scope does not see into the physical body, as other medical implements may. Rather it is thought to see into the interior of personality because it crosses a boundary that separates public fixity from private potentiality. The photograph presented an acute threat of this sort of invasion because it seemed to capture a moment out of the ambiguous flow of life—or an expression of personality that might have otherwise gone unnoticed—and fix it permanently in a medium epistemically prized for its objectivity.

We should not understand this unexpressed potentiality of personality in terms of its being unknown, as a secret might be, but rather *unknowable*, at least in the propositional sense that an object of knowledge must be either x or not-x. It is unknowable not because it is deeply hidden or well protected, but because it has not yet formed into a definite expression. In contrast with what is known *and* what is unknown (that is, undiscovered), the unknowable is essentially ambiguous, unformed, potential. This means that the realm of personality's potential must be obscure not just to others but to oneself, too: not hidden from oneself, but unknowable because still unformed. Unknowable and unformed does not mean unreal, however. We can become acquainted with these penumbral regions of potentiality by, among other things, the very experience of an idea or action coming to us "out of nowhere," as well as by bumping up against our own internal limitations of penetration and self-knowledge.

This is a much deeper invasion than simply trespassing into the domestic realm or slandering someone's good name, which explains why Warren and Brandeis would describe the harm against which privacy protects as "mental" and "spiritual" rather than "material" or "bodily."[60] They were far from alone in using the language of spirituality and the sacred to describe privacy. Their use of "inviolate" to describe personality, meaning not impregnable but sacred, draws on a long tradition of associating the sacred with concealment, and it echoes a common way of speaking about privacy at the end of the nineteenth century (recall "The Sacred Privacy of Home"), which continues to this day.[61] The inner mysteries of the sanctum sanctorum are not necessarily secrets; even when the architecture of sacred spaces is designed around hidden scrolls or tablets, it is not typically unknown or secret what is written on them. Rather, the inner sanctum's value as a sacred space comes simply from its being unseen (except perhaps by a select few, who are needed to testify to its existence). The sacred is *set aside*, in the literal meaning of the world, and its power depends on its ambiguous existence in the minds of believers. If a commoner crosses the threshold of a sacred space, he has perhaps trespassed in the temple edifice, but he has also invaded, and perhaps destroyed, the mystery of that realm by gaining knowledge of it as being one way rather than another. The same goes for privacy. Here again are the words of Warren and Brandeis:

> A man records in a letter to his son, or in his diary, that he did not dine with his wife on a certain day. No one into whose hands those papers fall could publish them to the world, even if possession of the documents had been obtained rightfully; and the prohibition would not be confined to the publication of a copy of the letter itself, or of the diary entry; the restraint extends also to a publication of the contents. What is the thing which is protected? Surely, not the intellectual act of recording the fact that the husband did not dine with

his wife, but that fact itself. It is not the intellectual product, but the domestic occurrence.[62]

This passage appears amid an extended argument in which the authors are at pains to dissociate the value of privacy from that of property, both physical and intellectual. But they go farther than that. What privacy protects in this case is neither information nor anything deliberately concealed but the private nature of the occurrence. This seeming tautology can be resolved if we think that the socio-ontological status of the man's dinner changes in a significant way depending on whether it is a private or public occurrence.[63] Once the event is publicized, the character of the event itself is altered—even after the fact! As in the other cases we have discussed, the relevant change isn't that some piece of information goes from being known by a privileged few (or one) to a wider audience, but rather that there is some definite information where before, at least as far as the wider world was concerned, there was a range of possibilities concerning how the man had dined. There was mystery and ambiguity. It doesn't matter that it was of a particularly quotidian sort. The curtain is rent, and something is lost.

The idea that the harm of a privacy invasion at its most basic consists in the deprivation or solidification of potentiality explains why Warren and Brandeis would make the rather unusual claim that "the principle which protects personal writings and all other personal productions, not against theft and physical appropriation, but against publication in any form, is in reality not the principle of private property, but that of an inviolate personality."[64] The analogy they draw between drafts of artistic works and human personality is meant to emphasize the potentiality of each: they are in essence unfinished, mutable, subject to change, but attain a kind of fixity when published. It is common to hear painters or novelists say that their works suddenly appear lifeless to them once completed, just as the potentiality of youth seems, in both experience and art, so much livelier when compared to the increasing path-dependence of later life. The existence of this potentiality is an element of well-being because it is, on this view, foundational to what it means to be a human.

The experience of such potentiality is also an element of psychological well-being for Warren and Brandeis. The solitary diner reaps psychological benefits from privacy because of "the peace of mind or the relief afforded by the ability to prevent any publication at all."[65] The peace of mind the diner gets from privacy is the same, they argue, as that of an artist who can "prevent the publication of manuscripts or works of art" that have not yet attained the fixity that comes with publication.[66] What sort of peace of mind can this be other than the reassuring openness of potentiality—the deeply sustaining confidence that one's life at its most basic and most mundane, like a work of art still in drafts or in sketches, is fundamentally open, mutable, and full of potential to be otherwise? One way of understanding this benefit derives from the Romantic view described above, but there are others. In the remainder of this book, I will give several different sorts of arguments for the benefit of this "peace of mind" that comes from being *oblivious* of publicity, none of which depend on the Romantic idea of the person. Indeed, the persistence of the idea that the human person is characterized by, among whatever else, a quality of potentiality essentially opposed to knowledge and information—and the persistence of this idea as a cornerstone of thinking about privacy across the years among post-Romantics, anti-Romantics, and all other sorts—gives strong warrant for thinking that it is an important part of what makes human beings human and makes their lives go well.

Again, this way of thinking is no relic of the Victorian era but runs throughout the history of privacy concerns, from the initial outcry over photography to twenty-first century anxieties about advances in brain-scanning technology. For instance, in a recent book calling for new protections of "mental privacy" to guard against the emerging dangers of neurotechnology, Nita Farahany recounts the threat to mental privacy posed by the use of brain-scanning technology in criminal investigations.[67] Although scientific research casts sufficient doubt upon the reliability of the technology to discredit its use in the courtroom, that hardly affects the conviction that it would nevertheless constitute an invasion of privacy to have one's brain scanned to see what one really thought about something and whether the scan contradicts or confirms one's verbal testimony on the matter.[68] In other words, brain-scan technology may be bad at uncovering secrets but, nonetheless, still poses a threat to privacy. In an essay on Farahany's book, Sue Halpern echoes nineteenth-century privacy advocates in suggesting why this might be: "What happens when companies, an employer, school administrators, or the government have access to our thoughts—or what they interpret to be our thoughts—before they are *articulated* or shared?"[69]

Although nineteenth-century privacy advocates might not have "anticipated governments' use of neural data," as Farahany puts it, it is a mistake to suppose that "no one, not even Brandeis, anticipated that governments could one day tap directly into our minds, deciphering the emotions, sentiments, and even unuttered speech that they detect."[70] As we have seen, privacy advocates in the Victorian era were concerned with the prospect of technologies for reading minds that seemed heralded by the invention of photography. This is the "alarming possibility" of the doctor's new scope. What's more, Farahany misunderstands their view of what privacy was thought to protect regarding "emotions, sentiments, and even unuttered speech"—perhaps because she is, like the rest of us, captive to the ideology that naturalizes information ("neural data," as she puts it) as always already existing in the private realms of human life.[71]

Yet sentiments and "unuttered speech" are not typically already formed mental objects that can be accessed and deciphered.[72] It is strange and surely mistaken to think that unuttered speech has a sort of objective existence in our minds which can then be accessed by some brain-scanning technology, as if all one's sentences were just waiting around in one's head for their turn to be uttered rather than coming into existence at the moment and through the act of our speaking them. The same is true for sentiments, emotions, and thoughts yet unthought. The privacy that a brain-scanner destroys is the same that the camera was thought to invade: precisely that state of affairs in which unspoken words (or other expressions) exist in the form that is particular to them—a condition of unknown potentiality that is extinguished when converted into information. Hence the echo and insight in Halpern's description of the brain scanner *articulating* one's thoughts rather than simply accessing or reading them.

Half of my motivation for excavating this forgotten view of privacy is scholarly: I think we have fundamentally misunderstood the foundational moment and texts of privacy and, in the case of Warren and Brandeis, one of the most politically influential articles ever published in an academic journal. The other half of my motivation is ethical and political, since there is value in this older view that we would do well to resuscitate. I have tried to avoid giving the impression that the originary ideas of modern privacy remain in the buttoned-up Victorian era, and I have noted several places where these old ideas underlie intuitions about privacy today, not least in the puzzle of the eye that invades. But they appear other places, too. Consider, for instance, the twentieth century's most famous text on photography, Roland Barthes's *Camera Lucida*. Barthes makes a claim similar to those advanced by privacy advocates a century prior: "Privacy is nothing but that zone of space, of time, where I am not an image, an object. It is my *political* right to be a subject which I must protect."[73] Barthes is not opposing privacy to the fact of appearing in a photograph per se but to the way that a photograph appears to reify the living multiplicity of a human being in a static object to be known; how the snapshot turns the fluid polysemy of life as it as lived (and as it dissipates into the flux of living memory) into something like information, if not information per se: fixed, static, manipulable, transmissible, and amenable to scrutiny. Barthes echoes the idiom of inviolate personality when he speaks of "the precious essence of my individuality" as "what I am, apart from any effigy."

> What I want, in short, is that my (mobile [photographic]) image, buffeted among a thousand shifting photographs, altering with situation and age, should always coincide with my (profound) "self"; but it is the contrary that must be said: "myself" never coincides with my image; for it is the image which is heavy, motionless, stubborn (which is why society sustains it) and "myself" which is light, divided, dispersed; like a bottle-imp, "myself" doesn't hold still, giggling in a jar. . . . Alas, I am doomed by (well-meaning) photography always to have an expression.[74]

Barthes's profound, private self is characterized by its slipperiness, its ambiguity and self-contradiction. The photograph fixes it as an image at the cost of its essential characteristics of versatility and potential. Tellingly, the photograph gives his face "an expression," which is at once a fixed look, a unit of meaning, and a legacy of the view of personality that associated spontaneous action with the production of authentic truth about the person. Barthes's identification with his slippery self recalls the Romantics, Emerson ("With consistency a great soul simply has nothing to do."[75]) and the rest of that still-ramifying lineage, but as Barthes shows, the view of the self as constituted in part by a valuable quality of unfixed, and therefore unknowable, potentiality is not the exclusive province of Romantic-inflected accounts. Nor is it an idea necessarily connected to the so-called liberal individual, negatively construed by its critics as a pre-political, metaphysically real agent that pre-exists its action and self-fashioning among, and connection to, the world of others.[76]

To take an example that will prove relevant later, we might consider the radical nominalism of Michel Foucault's post-humanist philosophy, which categorically rejects the existence of any pre-political or pre-social subjectivity but nevertheless repeats the idea about personality's essential quality of potentiality from a different point of departure: "My way of being no longer the same is, *by definition*, the most singular part of what I am."[77] Lest one get the impression that by being different Foucault means adopting a transgressive or nonconformist identity, he makes clear that *any* sort of fixity results in a similar loss. "The demand [*exigence*] for an identity and the injunction to break that identity both feel, in the same way, abusive."[78] It would seem to lend a high degree of plausibility to our arguments about the value of self-ambiguity and nonidentity, of spaces and protections of human potentiality against reification, that they play an important role in an enormously diverse array of philosophical frameworks deriving from radically different and often opposed foundational commitments concerning the metaphysical nature of the self, agency, society, and value.

For Barthes, like the nineteenth-century advocates we encountered, practices and rights of privacy are valuable in part because of the confidence they give individuals that certain aspects of their lives will unfold under conditions of oblivion, and in part for the actual potentiality that the oblivion of privacy protects and produces. The idea that privacy supports human flourishing by creating domains of life opposed to the production of information unites many an odd bedfellow: the statesman and the divine crawling around on all fours, ready to surprise themselves; Barthes and his view of the deep self as something protean, sprightly, and impossible to grasp or articulate without damage; Foucault's genealogical project concerned with the many ways that subjectivity is produced and disciplined by the manufacture and accumulation of information about individuals; and a pair of nineteenth-century Boston lawyers writing an article for the *Harvard Law Review*. Indeed, it is in this sense that Warren and Brandeis insist that we must not understand the value of the solitary diner's privacy as having to do with his ability to control who knows what about him: some control is necessary for his privacy, perhaps, but control itself is not implicated in their account of its moral worth. Rather, the diner's privacy is valuable because of "the peace of mind or the relief afforded by the ability to prevent any publication at all."[79] We need a legal right to privacy, they think, not just to prevent whatever harm stems from the violation of one's privacy (which they thought quite severe, "far greater than could be inflicted by mere bodily injury") but also to provide a social basis for the reasonable confidence that one can actually enter into oblivion from time to time, and that it will form a reliably available part of the structure of one's society.

Invasive Questions

I have only just begun to show how these ideas about the value of unarticulated potentiality at the core of the human person continue to animate our thinking about the value of privacy, yet already we should recognize that despite their vestigial presence in our moral intuitions, these ideas stand in tension with the standard defenses of privacy in our time, which tend to focus on the value of individuals' control over specific pieces of information or data.

Consider the invasive question. We recognize such questions as "invasive" less by their content, which can vary widely, than from the reaction they elicit. The natural response to an invasive question is "None of your business" or "That's private." Although invasive questions may involve matters of taboo, they need not. In the circles in which I move, it is thought to be invasive to ask someone how much money they make, whether they make love to their spouse, or even for whom they voted in the last election. What makes a question invasive will vary from culture to culture and person to person. This is because we tend to think it is up to individuals to decide what sort of questions are invasive and not, and for precisely the same reason we would object to the doctor's "meddling scope." Invasive questions are not invasive in virtue of their object being a secret. My wife and I have two unadopted children—it's no secret that we have probably done it once or twice. The response "that's private" to an invasive question about my sex life is importantly different from "it's a secret," "that's privileged information," or "I can't tell you that." The response to the invasive question is not a denial but a rebuke and the assertion of a boundary. That boundary protects against the creation of any information, accurate or not, concerning my sex life. The rebuke "that's private" or "that's none of your business" at once chides at the invasive questioner for attempting to replace the state of unarticulated ambiguity between us (at least regarding my sex life) with one of definite information, while also seeking to preserve the state of affairs between myself and my interlocutor that excludes any explicit information about my sex life. It is this boundary to which we refer when we call these questions invasive, and it is on account of the question's failure to cross it—by the success of the rebuke "that's private"—that we describe it as invasive rather than an invasion. Such questions attempt what the forced confession or surreptitious photograph achieve: getting the individual to reveal something about themselves. As with the forced confession, the question is not less invasive if it is asking after public knowledge or if I prevent the questioner from knowing the truth about me by responding with a lie.

If a colleague asks me whether I am intimate with my wife, they can guess the answer to that question; my reticence doesn't prevent them from knowing, although it does prevent them from *hearing it from me*. The question is invasive because I do not want to respond to it in *any* way, and my interlocutor has acted inappropriately by putting me in such a position. If you have ever been asked an invasive question, you will know that it feels like the other is *trying to get something out of you*, as the common phrase goes. Now we are in a better position to say from where and in what sense they are trying to get something out of you, and what that something is. On the one hand, the invasive question attempts to reach, as it were, into an individual's realm of interiority where agents make up their minds about who they are and what they will do. On the other hand, when my colleague asks me an invasive question, she isn't merely trying to know something about me, but is seeking to replace, in her mind and our relationship, a condition of ambiguous unknowing with a fact of the matter when, as my perception of the question as invasive indicates, I would prefer the former. This is why it does not make the question less invasive if it is one to which I do not know the answer, and why the natural response to an invasive question is not to lie or obfuscate, but to refuse to answer.

Although a lie protects a secret, it does not protect what privacy does, this condition of there being no information on the matter one way or the other—it does not protect oblivion, in other words. What I want is for the invasive question never to have been asked, or even better for the question never to have crossed the asker's mind. In other words, I want the asker to be oblivious, not mistaken or deceived as one would be with a lie. It is because we wish to guard this oblivion in the face of the question's attempt to pierce it that we do not respond to invasive questions with lies ("No, I don't have sex with my wife!") but by saying "That's private." And to the extent that our mannerisms, our blush or lowered eyes, seem to give an answer one way or another, we feel like we have failed to stave off the invasion. The lengths we go to guard our expression in response to the invasive question, to keep a straight face or puff ourselves up in indignation, are a legacy of our inherited beliefs and habits of perception about the telling expressiveness of the face.

It is time to return to the question with which we opened the chapter. Were Svenson's photographs invasions of privacy, as the court said, or were they perhaps merely invasive? The pictures in *The Neighbors* series are so effective because they evoke a large complex of charged ambivalences—they seem to teeter on the verge between private and public, respect and violation, knowable and unknowable, voyeurism and innocent looking—all converging upon the knot of social, ethical, epistemological, and political questions that we class broadly under the name privacy.

However, the most potent force in Svenson's photographs, and what makes them enduring works of art, derives from a calculated inner tension between the pictures' invasion of privacy's oblivion and their simultaneous reproduction of it. It is undeniable that Svenson took pictures of his neighbors without their knowledge or consent, which as we have seen is one of a few core examples of privacy invasion that link the origins of modern privacy to the concerns of the twenty-first century. Yet at the same time, the pictures Svenson chose to display, a fraction of those he had actually taken, are constructed in such a way as to preserve something like the privacy of their subjects. Of course, their privacy isn't literally preserved in a state equal to what it would have been had Svenson never picked up his camera. Nor does this seeming preservation have anything to do with the people in the pictures being unidentifiable or anonymous. Several of the subjects were in fact identified, and the faces of several women and a child can be seen.

Rather, the sense that the pictures reproduce something like privacy owes to Svenson's unexpected and often clever inclusion of the building's architectural elements within the frame of the photograph. Casements, exterior walls, and curtains, even window grime and glare, all conspire to conceal and, at the same time, reveal the neighbors' domestic space and bodies. Mixed in with the invasiveness of these images is confirmation that the viewer's eye can never penetrate the neighbors' lives totally. (This is, it is worth nothing, the opposite of the logic of surveillance.) The pictures in *The Neighbors* reproduce evidence of the unknowable within the uncovered and the known by gesturing, through their very act of exposure, toward domains of the neighbors' lives that lie unreachably beyond our powers of perception and knowledge. This produces a sense in the viewer that the lives of the individuals who appear in Svenson's pictures have depth and potentiality that necessarily exceeds our grasp, no matter how long we scrutinize the photographs. And if that is true about Svenson's neighbors in lower Manhattan, then it is true about the rest of us, as well—support, perhaps, for the artist's statement about portraying human universals in the particular lives of his subjects. This notion of depth, its vitality in human affairs, and its relation to oblivion and privacy is the subject of chapter 5, so for now let us simply mark that it is the reproduction of obstacles to the viewer's knowledge of his subjects that makes Svenson's pictures meaningful, lively, and mysterious. Above all, that is why they express something particularly *human* about the human beings who appear in them. Surveillance photographs, and pictures taken by paparazzi, cops, and private detectives—that is, pictures taken in order to serve as information—lack this quality, because they seek to portray human life in the static form of information, rather than as something whose quality of potentiality, surprise, and depth can never be exhausted because it is at some fundamental level unknowable.

2

PRIVACY, PERCEPTION,

AND AGENCY

In 1965, Gerald Foos bought the Manor House Motel in Aurora, Colorado, to satisfy his urge for voyeurism. He built special viewing platforms in the attic space above the rooms and installed false vents in the ceilings through which he could watch his guests undetected. He did all the work himself to prevent anyone from finding out. For nearly thirty years Gerald Foos spied on the unsuspecting guests of the Manor House; he was never discovered, and he never told anyone but his wife, who kept the secret. When Foos finally confessed his decades of voyeurism to the journalist Gay Talese, he insisted that he had done nothing wrong. According to Foos, he had not even invaded his guests' privacy. He figured that since he never got caught or communicated what he saw, no one was ever affected by his spying, much less harmed. And if no one's life was ever affected, then no one had any grievance against him. "A guest is entitled to his or her privacy," he said, but "there's no invasion of privacy if no one complains."[1]

Foos's moral logic is obviously flawed and self-serving, albeit far more common than one might think. It reappears in debates about privacy with unsettling frequency. For instance, in the wake of Edward Snowden and Glenn Greenwald's revelation of the NSA's widespread warrantless surveillance of US citizens, an interviewer asked Greenwald if he had evidence of anyone being *actually harmed* by the program—that is, in addition to the evidence that they were spied upon.[2] If this seems like one question too many, it is because we tend to think that to violate someone's privacy is to harm them in some way.

And yet it turns out to be rather hard to explain just how the lives of those guests and citizens would have been worse off had the violations of their privacy never come to light or been used in any way against them.

To get a better view of this difficulty, and to see how it points to an important puzzle at the heart of our concepts and practices of privacy, consider a somewhat simplified version of the Foos story.

Voyeur

Ben is a traveling salesman far from home who stops in at the Manor House Motel to spend the night. He rents one of the rooms where, unknown to him, Gerald Foos has installed an aperture for observing unsuspecting lodgers. Over the course of the evening, Foos sees Ben watch television, adjust his clothes in the mirror, and read a book. At some point, Ben goes into the bathroom and reemerges dressed in the sweatpants and t-shirt that he wears to bed and, in the morning, that he will also wear to the breakfast buffet. Foos is a simple voyeur and a total stranger to Ben. He never learns more about Ben than what he observes through the hole, and he never tells anyone what he saw. Ben never finds out about Foos's spying. Sometime later Foos forgets about Ben, and shortly thereafter takes the knowledge of his voyeurism to the grave.

Ben's story presents a puzzle for our understanding of privacy because we tend to assume two things to be true. First, it seems obvious that Foos violated Ben's privacy. However capacious our concept of privacy, no one can reasonably doubt that Ben's privacy was violated and that it was Foos who violated it by spying on him. Second, we take it for granted that it is a bad thing to have one's privacy violated. The assumption that privacy violations leave their victims worse off in some way is the basis for the idea that privacy rights get their moral force from the prevention of harm to the violated, rather than from an interest that potential privacy violators have in not sticking their noses into other people's business.

The puzzle arises because it is not easy to say just how Ben is worse off as a result of Foos's voyeurism. The spying does not affect Ben's mental states, projects, or relations with others. In fact, it is hard to see how Ben's life is affected at all. This has significant consequences for our understanding of the moral value of privacy, for if we are unable to identify a harm in Ben's case, then we are faced with a pair of unpalatable alternatives. Either we will have to give up on the first assumption and say that in Ben's case whatever interest privacy protects is unaffected, suggesting that we were wrong to think that Foos violated his privacy (and therefore that Ben has a moral right to privacy against such spying). Or we will have to abandon the second assumption and say that not all violations of one's privacy are harmful.

These alternatives are unpalatable because the stakes of abandoning either assumption are rather high. It is almost unthinkable that we would abandon the first assumption and accept the conclusion that Foos did not, in fact, violate Ben's privacy. The Peeping Tom is a central trope in our thinking about privacy, and we tend to imagine that we would have a grievance against Foos if we were in Ben's shoes. It would certainly come as a surprise to discover that we have misunderstood, so long and so deeply, what we care about when it comes to privacy. It is almost as difficult to imagine abandoning the second assumption. Ben has a grievance against Foos, we think, because Foos's spying affects Ben in some way that leaves him less well off. But if we cannot say how it is bad for Ben, or even how his life is affected, then we ought to consider that the reasons we have for supposing that Ben has a right to privacy are much weaker than we previously thought—perhaps even weaker than the reasons we have to permit all sorts of undiscovered snooping in name of public safety or other benefits to citizens and consumers. In a liberal polity reluctant to prohibit harmless behavior, it is always a short step from a judgment of harmlessness to "no harm, no foul." The notion that when it comes to privacy "what we do not know cannot hurt us" licenses a great deal of surveillance. Moreover, abandoning the second assumption would seem to generate moral reasons, based in the well-being of the spied-upon, for their spies to take pains never to be found out. It is rather perverse to think that Foos respects Ben's privacy interest by keeping his violation of it a secret.

For these reasons, let us take for granted that the ordinary view of the *Voyeur* scenario is correct: that Foos does violate Ben's privacy, and that it is bad for Ben that it happened. The rest of this chapter will be concerned with understanding what sort of support there may be for such a view, and what a justified account of the ordinary view reveals about privacy. For it turns out that understanding this puzzle does, in fact, reveal quite a lot about the moral value of privacy, much of which has gone unnoticed or unappreciated. To be sure, our explanation of the ordinary view of *Voyeur* might not explain everything that we talk about when we talk about privacy. More of that will fall to the remainder of this book. However, whatever else a theory of privacy does include, it ought to include an explanation of the ordinary view of *Voyeur*.

Privacy Harm and Well-Being

To call the violation of Ben's privacy harmful or harmless is to invoke a specific sense in which things can be bad for someone. A person is harmed, I assume, when something happens to them that injures, thwarts, or interferes with some interest, understood as a general component of human well-being.[3] There are a variety of ways to understand why it is *wrong* for Foos to violate Ben's privacy—and on certain views of moral character, how it is bad for Foos as well as wrong—but the general idea behind the moral right to privacy is that it is like the right not to be hit. That right rests on the idea that we have an interest in being in an un-punched condition because that condition (however we understand it) is an aspect of human well-being. By the same token, Ben's moral right to privacy protects an interest in well-being that is thwarted by the class of actions against which the right protects. This is why it is no excuse for Foos to say that Ben was none the wiser.

If we are right to assume that the violation of Ben's privacy is bad for him, it will obviously not be because of the variety of harms that can follow from privacy violations, for none of these obtain. It is important to note that this applies not only to the specific case of *Voyeur*, but to privacy more broadly. Consider the idea that what makes some act a privacy violation does not depend on the content of what the violator perceives. If someone peeps into your hotel room or reads your diary, it does not matter that what they observe is already public information or something they already knew. The substance of what is observed may bear on whether you suffer secondary harms in the domains of reputation, control over self-fashioning, the ability to present different self-images in different contexts, and so on. However, none of these harms follows necessarily from the privacy violation itself.[4] As I have already mentioned, locating the harm to Ben in anything other than the act of the violation itself also leads to perverse consequences, for it would seem to generate an interest in Ben that he never learn what happened, and that Foos guard his secret and take care never to be found out. It would be strange, to say the least, if the interest protected by Ben's right to privacy were also protected by Foos getting away with the violation of that right.

There is much in life that is bad without being harmful. Although it may be a bad thing to come down with a stomach bug or to get dumped, we do not tend to think of these experiences as harmful. This is because harms affect interests, understood as fundamental elements of a person's well-being. One surely has an interest in being healthy, but it is unreasonable to suppose that such an interest is injured or thwarted by a short-term illness from which one soon recovers. The same can be said for a great range of life's aggravations, disappointments, and discomforts. However in *Voyeur*, if it is bad at all for Ben that his privacy was violated, it cannot be bad for him in the way that a stomach bug would be. Since Ben never discovers what Foos did, and since whatever consequences there will ever be are confined to the time and place of Foos's spying, the violation of Ben's privacy will not be bad for him in any of the ways that things can be unpleasant, undesired, or hurtful. Rather, it must have to do with an alteration of a state of affairs that (1) can be said to affect Ben, (2) in a domain of well-being sufficiently important for us to impose duties on others not to interfere with it. The rest of this chapter considers these questions in turn.

Before going any further, however, we should ask whether I have confused a token of a privacy violation for its type. Perhaps we can admit that what happens to Ben is a token of the type "privacy violation" without conceding that this specific token (*Voyeur*) shares all the features of the type (specifically those having to do with harm).[5] Put differently, it might be that Foos violates Ben's moral right to privacy, which is a specific instance of a more general right that protects against the sort of harms that typically attend privacy violations but need not in every case. What happened to Ben is the exception that proves the rule, in other words. On this view, we could admit that Foos wronged Ben without having to concede that he did him any harm, even if we might still worry about the unpalatable consequences I raised earlier. This would be a mistake.

One reason to dismiss this objection is that it asks us to understand *Voyeur* as a special case, and therefore that Ben's right to privacy is justified not by the facts of his case, but indirectly, in virtue of certain similarities between Foos's spying to other, more central cases of privacy violations. But this cannot be right. Just as the Peeping Tom is central to our understanding of privacy, so too is the idea that it is irrelevant for privacy what Tom sees, only that he sees it under certain conditions. Another reason to resist the objection builds on the idea that exceptional cases are exceptional because of their relatively rare appearance in the world, and not on account of the conceptual difficulties they present. I find it reasonable to suppose that undiscovered privacy violations are more or less as common as those we uncover. In fact, given the technological sophistication of digital snoopers relative to that of the average computer user, I suppose that violations of this sort are, or at least very well could be, more common than those we discover. If that is at least plausible, then we have reason to think that *Voyeur* is not only one central kind of privacy case, but also that part of what counts for its centrality is precisely its going undiscovered. A theory that makes an exception of this case would therefore appear somewhat inadequate. Our sense that Ben would be better off without Foos's violation of his privacy does not rest on the exceptional application of a general moral principle. Rather, it expresses the judgement that Ben's life would have gone better had this not happened.

So, if the violation of Ben's privacy in *Voyeur* is bad for him, it must be because it adversely affects some interest of his. And if he has an interest that is capable of being affected by Foos's spying, then it must be an interest that Ben has in a state of affairs independent of his mental states, life plans, and relations with others. This is significant because it means that underlying the ordinary view of *Voyeur* is an assumption that Ben's interest corresponds to an externalist view of his well-being. The basic idea of the externalist view is that there are at least some elements of a person's life that can go better or worse independent of that person's awareness of them. This is a venerable thought whose best-known modern articulation is probably Robert Nozick's experience machine.[6] What the experience machine was supposed to show is that for certain aspects of human life, what counts for well-being is a state of affairs actually obtaining, not merely one's having the corresponding psychological experience. The ordinary view of *Voyeur* expresses the thought that what we want for Ben, and therefore ourselves, is not just to experience privacy, but actually to have it. This is significant because it suggests, against a strong current of contemporary thinking about privacy, that to some extent we care about privacy for its own sake, and not just because it has instrumental value relative to other things that are important to us, like autonomy, reputation, and material gain.

If you are skeptical, consider the choice between a world in which distant aliens, who were somehow unable or prohibited to make contact with Earth, watched you in the bathroom and the bedroom, and a world where they did not. Or instead of aliens, perhaps a hacker on the other side of the world who keeps her spying to herself. My guess is that in neither case would we consider the choice to be a matter of indifference or properly subject to non-moral judgments, like matters of taste. Rather, I think most would choose the world without the watchful aliens, and that if pressed to explain our choice we would say that with the aliens we would be worse off in terms of privacy— that is, in terms of whatever it is that privacy protects—even if we still cannot say exactly what we mean by that.[7]

Perception and Deprivation

The first challenge in explaining how Ben could be harmed by the violation of his privacy, even in the externalist sense, is to explain how Ben is affected at all by Foos's peeping. For Foos to harm Ben, it is obviously necessary that Foos affect his life in *some* way, and it is not yet clear that Foos has, in fact, affected Ben's life simply by looking at him. Compare the privacy violation in *Voyeur* to a pair of other undiscovered wrongs, one in which the question of harm never arises, and another where the description of harmless wrongdoing seems fitting.

> *Undiscovered Assault*
>
> X takes pills for insomnia that deliver eight hours of the most profound, unrousable sleep. One night, Y enters X's bedroom with the intent to assault X. Sometime later Y leaves, having left no trace on X's body or anywhere else, and never tells a soul.

> *Undiscovered Trespass*
>
> Again X is asleep. Y enters X's house through the unlocked door, but instead of going into X's bedroom, Y explores a bit, looks around, disturbs nothing, and then slips back out, leaving everything exactly as it was. Again, no one but Y ever knows what happened.[8]

In all three cases, the consequences that ensue are irrelevant to the question of what makes the act wrongful. However, *Assault* differs from *Trespass*— and resembles the ordinary view of *Voyeur*—in that there is never any question about harm, either. Whereas it is conceivable to say that the trespasser wronged X but caused no harm, Y's successful assault would be ipso facto harmful for X. If Y does not harm X in *Assault*, then we are bound to assume that once inside the room, Y had a change of heart and did not in fact assault X. If Y does assault X, then it might additionally be bad or harmful if Y tells others what happened—X might feel humiliated or ashamed or might simply not want to be known by anyone as the victim of an assault—but these sorts of secondary harms bear no necessary relation to the harm of being assaulted, nor will they answer the basic question of why it is harmful for X to be assaulted in the first place. The same would appear to be true of *Voyeur*. Indeed, the sorts of ancillary harms that can follow from an assault are the same ones that privacy is often supposed to guard against: injuries to social standing, self-image, one's relationship with others, the ability to present different self-images in different contexts, and so on.[9] These are bad outcomes, to be sure, but their relation to privacy is no more necessary than it is to assault.

I think that *Assault* and *Voyeur* strike us as similar, at least when set against *Trespass*, because they are instances of one person doing something to another. By contrast, the most natural way to describe what happens in *Trespass* is that the Y does something to X's property (enters his house, looks at his stuff, etc.). Of course, inanimate objects are not the sorts of things that we typically understand to have interests, and therefore a trespass will be harmful to some person or it will not be harmful at all. What I mean to pick out by contrasting *Trespass* with *Voyeur* and *Assault* is the kind of violation whose object is the person itself.

Yet the idea that the object of Foos's privacy violation is Ben's person is just what accounts for its air of paradox: Ben seems at once touched and untouched by Foos's gaze. We can get clearer about this by distinguishing *Voyeur* from *Assault*. Note that different aspects of the victim's person are affected in either case. Whereas the object of assault is the physical body, the object of privacy violations is not, or at least not in the same way. Consider a variant of *Voyeur*.

Hacker

A hacker develops a computer program to infiltrate computers across the globe at random, while also making it impossible for the hacker to identify or contact their owners. The program gives the hacker access to a person's emails, photos, documents, and so on. Like Foos, the Hacker is a simple voyeur who never tells anyone what he sees, and no one ever finds out.

As in *Voyeur* and *Assault*, it is natural to say that if the hacker were to target you, he does something *to you* rather than to your computer, although technically the opposite is true. This is because the object of privacy violations is not the physical body or domestic space but the epistemic dimensions of personhood—that is, those aspects of a human life about which knowledge can be produced. (Forgive the awkward locution, but it is important not to commit the common error of taking for granted that the protections of privacy concern aspects of human life that are not just knowable, but are so because they already consist of information.) This is true in *Voyeur* as well as in *Hacker*, since it would be more accurate to say that the object of Foos's peeping is Ben's *appearance*, rather than his physical body. If Foos were to come into unwanted physical, rather than merely perceptual, contact with Ben's body, we would probably say that a line had been crossed from privacy violation to assault. This is also why it may appear to be a greater violation of one's privacy if someone were to break into one's home rather than, say, an apartment one owns but rents out: one would probably feel that the former, by being the place one keeps all one's stuff, is more closely connected with oneself than the latter. By the same token, *Voyeur* and *Hacker* differ from *Assault* in the mode of violation, which in the privacy cases might be described as learning something about the victim, or coming to know them in a particular way.

Let us call the aspect of the person about which we can learn without asking the person him- or herself *the biographical dimension* of that person's life.[10] Ben's biographical dimension includes what can be said about his physical appearance and behavior, among much else, and it is not limited to what can be learned firsthand. Much of it will consist in secondhand information: facts about him, commentaries on his appearance and behavior, and second-order normative statements. The biographical dimension of one's life is made up of beliefs about what one is like and has done, which can exist in the minds of others (as in *Voyeur*) as well as in other, more permanent forms like writing and photographs (as in *Hacker* and actual biographies).

What is most important here is that the body of beliefs about what Ben is like, now and in the past, is not merely descriptive but also constitutive of who Ben is. The beliefs comprise this part of him. For example, whether Ben is generous or a dancer will depend on whether he does, in fact, give freely or dance in public. Still, it is possible that no one takes him seriously. Maybe to all the world his giving is tainted by ulterior motives, his dancing unserious and ungraceful, such that the answer to the question of whether Ben is generous or a dancer will be "no" or "not really." Although this aspect of Ben is external to his mental states and physical body, we cannot really say that it is therefore separate from him, or that it is less a part of his life than his physical body. For if the only qualities of Ben's that we could call integral to him consisted in properties independent of his relation to others and what they thought of him, then we would find ourselves with very little to say about him. We could describe his physical organism, and his location in space and time, but not any of the elements that make him an individual in a robust sense, nothing that gives his life color, direction, and a sense of meaning. Without reference to his biographical dimension, Ben no longer appears like an individual at all, but instead merely a member of a species.

Here is a sense in which Ben is not unaffected by Foos's undiscovered voyeurism. By observing Ben, Foos formed part of his biographical dimension. If part of Ben is constituted by the independent perceptions and evaluations of others, then that part of him can be affected independently of any effect on his mental states, life projects, or relations with others.[11] This brings us closer to understanding how Ben could be harmed by Foos's spying, but we still have not succeeded, since the impression that Foos forms of Ben is identical to aspects of his biographical dimension that preexisted Foos's snooping. It is not the case that Ben is harmed because what Foos sees reflects negatively on him or is out of keeping with how he wants to be seen or comports himself in public. We might even suppose that Foos leaves his hidey-hole overwhelmed with admiration, without affecting the intuition that it was nevertheless bad for Ben that his privacy was violated. It is hard to see how Foos alters Ben's biographical dimension at all except in the sense of forming an impression of Ben where there had been none before.

We can now give a provisional and partial account of the harm in *Voyeur*: Foos harms Ben not simply by forming a part of his biographical dimension, but by depriving him of a degree of possibility in that dimension that would have existed but for Foos's spying. In other words, the harm does not consist in *what* Foos makes of Ben, since what he makes of him is no different from the impression he would have formed at check-in or the continental breakfast. Rather, by forming any impression at all, Foos deprives Ben of something he otherwise would have had. Understanding *Voyeur* as an instance of deprivation, rather than one of infliction or interference, fits with a common way of describing Foos's action: he deprives Ben of his privacy. When Foos arrives at the peephole, Ben is suddenly and for that reason without the privacy he would have otherwise enjoyed—which, on this provisional view, means that Ben has been deprived of a state of affairs in which, for the time being, there was no biographical information about what he was like. Before Foos sees Ben standing at the mirror and sleeping in sweats, the man could have had any number of characteristics for him—not infinitely many, but the range of possibility is vast.

I admit that this is an unusual way to understand the harm of *Voyeur*. In a moment I will move beyond this explanation, but first we must note a few of its interesting points as a theory of privacy. First, it fits with the origins of privacy that we excavated in chapter 1. The idea there was that the modern value of privacy originated in a view of the self according to which "something is lost, not exhibited or gained, when infinite possibility gives way to the limitations of actuality,"[12] an idea that had its roots in Romantic thought but also plays an important role in post- and even anti-Romantic ideas about the self. Next, it makes good on our sense that Foos does something to Ben as a particular individual and not as an abstract bearer of rights or dignity. The discomfort we feel at being watched ("his eyes were all over me . . ."), and which we may feel vicariously for Ben, resembles the repulsion from an unwanted touch. We can understand this if we concede that a part of who we are exists beyond our physical bodies and capacity for self-government, specifically in the minds of others, and therefore can be affected or "touched" by the simple act of another taking us to be a certain way. A leer or sexualized gaze on the street makes us uncomfortable not because we believe that a gaze can actually touch the physical body, as the metaphor goes, but rather because we imagine the ogler's dirty thoughts to affect, however minutely, the parts of our selves that exist beyond the bounds of our bodies and minds. The sexualized gaze seems to make contact because it gives us the feeling that the ogler really is forming us in a certain (unwanted) way in his mind. This is what we mean when we say "what do you make of him?" to ask what someone is like.

Further, this view offers an explanation for why it does not matter for privacy what Foos sees, only the circumstances under which he sees it, since the one who violates another's privacy deprives him of some possibility even if what the violator observes is no different from what anyone can see in public. This gives us reason to think, as I expect we do, that it would be more harmful (rather than less or equal) if Foos spied on Ben for a week straight without seeing any change in Ben's appearance and behavior. Indeed, one advantage of this account of privacy over those based in the value of maintaining a public image, reputation, or freedom in self-fashioning, is that it has no trouble explaining the privacy interest of one who is the same in both public and private. Against both the argument that "privacy is only for those who have something to hide" as well as the feeble response that "everybody has something to hide," this view draws our attention to an interest and source of value in human life that everyone has simply by virtue of being human and which is lost even in the violation of the privacy of someone whose public and private comportment are completely consistent with one another. The important point is not that there are many such people—though how would we know?—but that the idea of privacy as a fundamental right or interest is inconsistent with the possibility of there being certain persons to whom it cannot be said to apply because of how they live their lives.

This account of privacy also helps to explain why we do not think that the animals who peek into our windows—the cats and the birds and the squirrels—thereby violate our privacy. Suppose that a racoon breaks into the motel attic and stares down through the peephole at Ben. Surely we do not think that the animal has violated his privacy. It isn't because we tend to think that nonhuman animals aren't fitting objects for questions of blame and moral responsibility; what interests us in the first place is whether a raccoon is *capable* of doing whatever it is that Foos does to Ben, which is separate from, and antecedent to, the question of blaming the raccoon or holding it morally responsible for looking through the vent. Nor is it that a raccoon is incapable of communicating what it sees—what would be the harm to Ben in that, anyway? Rather, the judgment that the raccoon does not violate Ben's privacy is based on the assumption that raccoons do not draw inferences about what a person is like from a series of observed attributes. If raccoons spoke English but had no concept of a person's attributes contributing to an overall picture of personal identity or character, then I suspect we would still think them incapable of violating our privacy. In such a world we might worry, in terms of privacy, that an English-speaking raccoon could observe us through the window and communicate the bare facts of his observation to someone, and therefore we would take care to draw the blinds. But this does not mean that the raccoon is capable of violating my privacy any more than a camera is.

What seems to matter for privacy is that the information make it to a human (or similar) intelligence that is capable of understanding that information as not just emanating from me as light from a star, but being about me in a deeper sense of my biographical dimension. If one day we discover that nonhuman animals do, in fact, attribute characteristics to persons in this way, then I expect that we would eventually come to think them capable of violating our privacy *even if* they lack human language. (If you discovered a man spying on you in your hotel room, but then learned that he was a foreign tourist with no knowledge of English, perhaps even the last speaker of his language, I expect you would hardly feel relieved. And I doubt that you would revise your opinion about whether your privacy had been violated or not.) Thus the uncanniness of being observed by certain animals, like raccoons and dogs. The sense of the uncanny comes from the similarity of the animal's eyes to human eyes, and of the intelligence flashing there to ours, while all the while remaining recognizably nonhuman. This also helps to explain our unease regarding the algorithms that compile profiles of us by tracking our Internet usage. These algorithms are not intelligent in the way that humans or even racoons are, but they are designed to mimic precisely that part of intelligence that is of concern here: the imputation of observed attributes to an underlying, unified self.[13]

This view casts light on yet another puzzle about privacy that, to my surprise, seems to go unremarked. Despite disagreements about privacy's value, more or less everyone thinks that, as a descriptive matter, what privacy *does* is protect against the sensory perception of oneself and an associated range of information and objects. And yet our imagination of privacy violations is limited almost entirely to the visual and the auditory. Why shouldn't it violate Ben's privacy for Foos to smell him, or taste him, or touch? A theory of privacy ought to be able to explain why violations have to do only or primarily with sight and hearing. The answer cannot be that touch, smell, and taste give no information about a person, or that those senses are less refined than the others. Nor can it be that being close enough to touch, or taste, or smell someone already puts one in such close proximity so as to have obviated any possibility for privacy, since this begs the question of privacy as a protection against only sight and hearing. A view of privacy as concerned with possibility in one's biographical dimension can make sense of this, for there are not that many ways a person can feel, smell, or taste. We more or less know in advance what it would be like to touch the skin of another (hence the agony of anticipation), and the ways that a body can taste or smell are fairly limited and common among human beings. Anyone with a body already knows what it's like. More specifically, it is not that one learns nothing

new about someone by touching or tasting them—Foos need not learn anything new about Ben to violate his privacy—but that there simply is nothing
new there to learn.[14] When it comes to smell, taste, and touch, there just is
not very much possibility to deprive. By contrast, in *Voyeur*, although Foos
learns nothing about Ben that he could not have learned from seeing him in
public, he learns it in a domain characterized by a vast range of possibilities
for people to be different from one another and themselves.[15]

Finally, the discussion so far calls our attention to a glaring error in the
standard accounts of privacy, which is that they beg the question that privacy,
even as a descriptive matter, is concerned with the protection of information.
It is clear that what results from a privacy violation is that someone comes
away with information that he did not have before—even if it is just Foos
knowing that Ben is the same in his motel room as he is outside of it. But it
is a mistake to assume that because information results from a privacy violation, information is therefore what privacy protected before the violation
occurred. It still needs to be demonstrated that what privacy protected before
its violation was the information thereby gained (again, this is what secrecy
does). To understand what we get from privacy, as opposed to what we get
from its violation, we might consider an example of someone who actually
has privacy—from us, that is, the readers of a text or viewers of a film—rather
than someone whose privacy has already been compromised. This is harder
to do than one might imagine. Such an example cannot take the form of
Ben's story, or of a person in the shower, or a picture in a safe, or really any
of the examples given in the literature on privacy, for such a person does not,
as far as we the readers are concerned, ever have privacy to begin with. It is
rather amazing how hard pressed one is to find a single example of privacy,
as opposed to its violation, in all the history of philosophical thought on the
subject. An example of privacy would have to demonstrate something like
what Joshua Rothman identified as the quality of impenetrability and mystery that Virginia Woolf lent to her characters. If we had constructed Ben's
example as an example of privacy instead of its violation, then we would not
have known what we did not know about him. We wouldn't have been able
to say how he was dressed, how he acted, or really anything about him while
he was in his room. We would have been oblivious, in other words. Only
then would Ben have what privacy gives: oblivion for us and, as I will put it
in a moment, unaccountability for himself. The best way of understanding
what happens in the violation or invasion of privacy is not that some reified

piece of information passes like a chit of property from its rightful controller to someone else, with or without consent, but that a piece of information is created where there had been none before.

Agency and the Unaccountable

The problem with our provisional explanation is that it offers a necessary but insufficient condition for the harm in *Voyeur*. For example, if we changed the setting of *Voyeur* to a city street but held everything else the same, there would be no question of Foos violating Ben's privacy, even though he did form a bit of his biographical dimension where it had not existed before. To explain how it could be harmful for Foos to deprive Ben of some possibility in the motel room, we will have to say more about what distinguishes the room from the avenue, and why Ben has an interest in being unobserved in one but not the other. My claim is going to be that the difference is that the motel room's reliable barriers against perception protect and enable a state of affairs in which Ben can be unaccountable. I will explain what I mean by that phrase and give some reasons to think that it is the sort of thing in which Ben has an interest, but first let me begin by offering a slightly different angle on *Voyeur*.

Another way of describing what Foos does to Ben is that he alters Ben's condition from one of privacy to one of publicity. Once altered, Ben's new condition can be said to have certain features that distinguish it from his previous condition of privacy—features that are independent his awareness of them. The most obvious of these is that being in a condition of publicity means that Ben has a public or audience where he did not before. To be more precise, it is the arrival of a public (of one) that puts in him in such a condition. When Foos mounts the peephole, he puts Ben on stage, as it were, and it is irrelevant for Ben's being on stage that he is aware of it.[16] It is also irrelevant to the question of whether one has an audience how large that audience is—a single person is enough.

One difference between Ben's new condition and his previous one is that having an audience (being "on stage") gives him reason to consider the fact of his publicity, specifically that he appears for others. In a basic sense, this is simply what it means to be self-aware. Walking by a precipice gives Ben reason to be concerned with his balance, and being onstage gives him reason to take account of that fact. These are reasons that respond to features of Ben's situation, independent of his mental states. The difference between being on stage and near a precipice (or the stage edge) is that being on stage gives reason to be aware of appearing before others. Since it is reasonable to expect that anyone who sees us will interpolate from what they perceive to a judgment about what we are like; and since such judgment, albeit provisional, comprises a part of who we are, we have reason to concern ourselves with the fact of our publicity to the extent that it matters to us how we are.

Ben has reason to concern himself with his publicity because he is accountable for himself in two related ways important to his agency and the belief that his life is his to direct. The first is connected to the observation that it matters to Ben how he is. To some (morally significant) extent, it is up to Ben what sort of person he turns out to be. Accounting for oneself means taking a stand on who one is and putting the matter one way or the other—not once, of course, but frequently over the course of a lifetime. We cannot will ourselves to be any way we choose or create ourselves ex nihilo, but it is a basic assumption of agency and moral responsibility that we are in a position to make up our minds about the sorts of people we shall be and then guide our actions accordingly. And for Ben to be accountable *for* himself in this sense, he must be accountable *to* himself. That is, in order to direct his life and understand it as being *his* in any meaningful sense, he must be able to distance himself from the manifold of empirically true things that can be said about him and decide which of them he affirms and which he rejects.[17]

The second way that Ben is accountable for himself is expressed by the idea that others will hold him accountable for how he acts and appears to be. This includes ideas of moral responsibility and being "called to account," but it is broader than that. What I mean to invoke here is how, by perceiving Ben *as a person*, Foos assumes a certain relationship between Ben's appearance and behavior, on the one hand, and who he is in the deeper sense of personality or character, on the other. This is just what makes characteristics characteristic.[18]

Of course, being accountable for himself in these ways gives Ben reason to consider how he appears even when he is alone and in private. The difference is that in private the activity of self-consideration presents a different kind of necessity. When one appears in public, the consideration of the fact of one's appearance is a practical necessity owing to the features of the situation. By contrast, reflecting on how one appears and acts in the world—if only to compare with how one desires or aims to be—is a conceptual necessity.

There are plenty of theories of just what sort of conceptual necessity this is, but whether it is something you have to do in order to be a person, or an agent, or to enjoy freedom of the will, in any case it is not something that you have to do all the time.[19] In fact, someone who never stopped reflecting on how they were would appear to be deeply narcissistic, perhaps crippled by self-reflection. The narcissist is overly focused on accounting for his features; the one crippled by self-reflection, like a person obsessed with ethically clean hands, undermines their capacity for action by never letting themselves off the hook. To either of them we might say, "You need to get out of your head," in an attempt to express the view that there are areas of life and ways of being with ourselves and others where it is worthwhile and even necessary to let go of self-consideration. Note that this is not like shifting one's attention from one activity to another, say, from revising a troublesome essay to cooking dinner. Nor does the injunction to get out of one's head refer to a transcendent self capable of escaping the "head" of its thoughts for some "outside" where it can be apart from them. Rather, the recommendation to get out of one's head or let oneself go means to ease up for a moment the insistence on accounting for oneself. The advice is to come apart, in other words.[20]

This coming apart plays an important role in agency, in addition to the psychological and existential benefits of "letting go" from time to time. A remarkably broad range of philosophical views of agency and what it means to freely direct one's life recognizes that the ability to separate oneself from the constitutive commitments of one's self-conception—to come apart in terms of self-knowledge and personal identity—is a necessary condition for saying that those commitments are one's own to begin with.[21] For instance, say I grew up in a politically conservative household. For a long time my understanding of myself and the world is the one I inherited from my parents and their milieu. But then one day, I come to reflect on those beliefs and ask myself if they are the sort of things that *I* really believe, that *I* endorse as fit to guide my life, or whether they clash with other beliefs and commitments I have picked up from other sources and experiences along the way. This is a process that tends to begin around adolescence, which jibes with our sense that such self-reflection is part of being a mature person, whose autonomy we feel bound to respect because, in part, her life is *hers* in just this sense. No matter whether one answers yes or no to these reflective questions of self-conception, it is the reflective stance of self-separation itself that leads us to believe that the way we live our lives (even if in conformity with how we were raised) is our own, especially when compared to someone who goes through life without ever engaging in such an experience of self-reflection.

Yet, as Charles Taylor notes, this example seems to involve the transcendent self stepping back and looking at *some* of its commitments while remaining fundamentally intact—how can we say such a person has really considered herself if the activity of self-reflection always leaves a part of the self (that is, the judging part, which is rather central to self-understanding) beyond the bounds of reflective interrogation? Taylor's answer is that we must be open to the "radical reevaluation" of our deepest beliefs and commitments. This is an act that must be done from a standpoint of nowhere, as it were, not against a moral "yardstick" but with reference to "my deepest unstructured sense of what is important, which is as yet inchoate and which I am trying to bring to definition. . . . To do this I am trying to open myself, use all of my deepest, unstructured sense of things in order to come to a new clarity."[22] This is an experience of letting oneself go completely, of being radically open and ambiguous to oneself in the absence of the self-knowledge that ordinarily structures one's understanding of who one is.

Taylor's use of "inchoate" to describe the domain of the self that we would expect to encounter in such an experience is highly apposite. The inchoate exists in a state of potentiality that has yet to assume a degree of fixity by which it can be known as one thing or another. What is inchoate is not imaginary or merely possible but has a verifiable existence, albeit of a protean sort that has not yet assumed the degree of fixity by which we can say that it is one thing or another. We can, as Taylor notes, come in contact with the inchoate, and we can become acquainted with it, although we cannot know it in the propositional way that we can know information, for to have attained the fixed qualities of information is to no longer be inchoate. Acquaintance with these unstructured, inchoate regions of the self offers a powerful resource for self-determination and the sense that one's life has depth and the capacity for change. Of course, for a person to be called an agent at all, she would need to come back together after such an experience of radical openness. It is because we emerge from this coming apart that "when we do it, we can be called responsible for ourselves."[23] There is no reason to think that this is an easy or everyday experience. Taylor himself recognizes that "the obstacles in the way of going deeper are legion."[24] Among such obstacles is the condition of publicity.

I want now to revise the provisional view by characterizing Ben's interest in *Voyeur* as an interest in being unaccountable. Being unaccountable combines the idea of possibility and ambiguity of the self expressed by the provisional view, while incorporating the idea that agency and well-being more broadly require opportunities for temporarily shrugging off the self-accounting necessary for self-knowledge and action.

We can begin to build out this view by recalling the idea that a person's biographical dimension is integral to who he or she is, to which we can add that it is not always easy to be the sort of creature about which this is true. I do not mean to say that it is challenging to end up as the person one aims to be, but that it takes sustained effort to be anyone at all. We have already noticed that much of who we are turns out to be essentially out of our hands. In addition, the bits that *are* up to us, or that are sensitive to our higher-order attitudes—like the desires, beliefs, and aspirations that give our lives direction and depth—are hardly self-sustaining. It is not enough to for Ben to be kind that he desires, intends, or understands himself to be so. He must actually act with kindness for it to be true that he is kind, or even that he desires to be kind in anything but the most trivial sense.[25] Even then, there is no guarantee that kindness will be reflected in his biographical dimension: that he will be kind in the more robust sense according to which we might say, after his death, that Ben was a kind soul. And yet it is a basic assumption of agency and moral responsibility that we are, to a significant extent, accountable for the sorts of persons we turn out to be. All this means that, for human agents, living a life takes persistent effort.[26] You have to keep it up if you would be anyone at all.

But this persistence, like any other form of striving, can be exhausting if unaccompanied by periods of repose. When there is no quarter from striving, persistence turns inward; self-direction becomes self-torment and depletes what it had hoped to sustain. Any picture of well-being ought to include, in addition to the importance of agency and moral responsibility, a space and time to shrug off the persistence that it takes to live a life. Such repose would not consist in the abandonment of one's ends and values, or caring about different things or nothing at all, but simply in not being definite, one way or another, for a little while. In other words, it would look like a period of not being accountable for oneself, to oneself, and to others. However, one way of being definite (or at least not indefinite) and of being accountable is to appear before others. When Foos mounts the peephole, Ben goes from not necessarily being anything for anyone, to definitely being something for someone, even if he does not know it. If this seems a bit like Ben working in his sleep, recall that it is Foos who is doing the work here. If repose consists in being unaccountable (to oneself and others), then Ben is not resting in the relevant sense while Foos is watching, even if he thinks that he is. This is not to say that Ben certainly would have been resting in this way had Foos never come to the peephole, but rather that Foos deprived Ben of the conditions that make such rest possible.

To my eye, someone who had no use for such repose would appear flat or one-dimensional in the same way that someone who never sets aside the labors of individuality strikes me as lacking in the depth that makes a person interesting. Such a person may turn out to be a monomaniacal genius, though more often than not they appear to us like over-strivers, deformed by their commitment to make themselves one way rather than another. And then there is the pathetic obverse of this character, who is perhaps uneasy with personality's manifold, and whose mania for recognition at any cost leaves them indifferent to the sort of recognition they get.[27] Both characters insist too much, we want to say—a phrase, like the line of thought I have been developing here, underwritten by a picture of well-being that includes a time and space for letting go. This is the aspect of well-being that the monomaniac sacrifices in the relentless pursuit of what appears to be of higher value. This is what, in the domain of privacy, those who live in the public eye sacrifice for fame. With both the monomaniac and the movie star we can point to an interest here, even in its sacrifice for something perceived as a higher good.

The value of this form of repose draws on the broader picture of well-being that underlies and animates the ordinary view of *Voyeur*. I will now bring that picture into sharper focus by asking whether being in a state of unaccountability is not only good in virtue of its relation to agential persistence, but also on its own, and whether privacy is a necessary condition (or something like it) for that state to obtain. The answer to both questions is yes, I think, though showing why will involve abstracting a bit from the facts of *Voyeur*. Nevertheless, the idea that what we want in *Voyeur* is not just to have the psychological experience of privacy, but for that state of affairs actually to obtain, will depend for its moral force on a broader view of well-being. The broader view I have in mind is, very roughly, one that features the usual slate of goods, including many that have to do with self-control and knowledge—freedom, autonomy, self-knowledge, friendship, and so on—but which also includes goods apparently opposed to these, goods like obscurity to oneself and others, and the enjoyment of the parts of human experience that exist beyond the limits of control, self-definition, and self-knowledge. I will now begin to sketch out what I take to be the most relevant and compelling elements of that view as they pertain to the discussion of agency. The rest of this book will be dedicated in large part to expanding, refining, and defending this view.

First is the idea that being unaccountable is not merely valuable for the role it plays in agency but is also a good in its own right. We could make this claim more specific by arguing that an aspect of well-being consists in having sufficient opportunity to confront ambiguous or contradictory domains of the self without attempting to resolve the situation and without it being resolved for us by others. But is this not to suggest an interest in the failure of agency? Someone who was unable to take a stance on themselves when faced with a contradiction or ambiguity about how they are would certainly appear to be, as Harry Frankfurt says, a "passive bystander to his desires and what he does."[28] This sort of passivity matters to Frankfurt and others because, in his words, the "difference between passivity and activity is at the heart of the fact that we exist as selves and agents and not merely as locales in which certain events happen to occur."[29] An important difference between the sort of creatures, like us, who live a life, and those that are merely alive is that "we are not prepared to accept ourselves just as we come."[30]

Yet the experience of taking ourselves just as we come only seems passive from the perspective of agency and practical reflection. There are other ways of being in the world and relating to oneself (and others) that are open and receptive, if not necessarily passive. I have in mind particularly the forms of receptive openness associated with the aesthetic and the erotic. The way we relate to a painting or a caress is not *merely* passive in Frankfurt's sense. Rather, the erotic and the aesthetic offer different ways of being with and knowing the object, be it a painting, a lover, or oneself.[31] These are modes of experience characterized in part by the pleasurable and enriching interplay of contradiction and obscurity, beyond the reach of self-control and self-knowledge. Accordingly, they tend to wither or recede in the light of practical reflection and self-awareness—especially when we consider the fact of our appearance. The most famous statement of this type of state is probably Keats's notion of negative capability: that "of being in uncertainties, mysteries, doubts, without any irritable reaching after fact and reason."[32] Just as we can relate to a painting or lover in this way, so too can we relate to ourselves. Taking oneself as one comes, with all one's manifold contradictions and obscurities, is at odds with being accountable for and to oneself in the sense of making up one's mind. But to the extent that there is more to our minds than what we make up, and that human beings are more than agents, the opportunity to be with ourselves in this way would appear to be an important part of human well-being. It is the sort of experience in which we would have an interest, in other words, which may in turn ground rights and duties.

If you have ever seen lovers kissing passionately in a public space—or if you have been such a lover yourself—you will know that their obliviousness to the world beyond their embrace creates a sort of temporary, mobile analog to the privacy of the motel room, such that they seem to enjoy what we have described as Ben's unaccountability, notwithstanding their doing so in public. The pseudo-privacy of reverie and ecstasy ("the world fell away" "I felt like I was the only one in the room") also suggests that actually being in private is not necessary for the psychological experience of being unaccountable, even if it might be highly beneficial. However, the externalist view of well-being behind *Voyeur* suggests that we should not understand the unaccountable only on the level of human psychology. Rather, the idea that Foos deprives Ben of a state of unaccountability that would have obtained in the absence of Foos's spying suggests that what we want is actually to inhabit a state like Keats's negative capability, and not only to have the corresponding (valuable) psychological experience. This is a view of well-being in which being unaccountable from time to time is a feature of a life well lived. In the ordinary course of things, inhabiting a state of unaccountability requires reliable material supports against perception (walls, blinds, etc.). And since not being perceived is a necessary condition of being unaccountable, we might consider others duty-bound not to interfere with those material supports.

There is yet another way that being unaccountable contributes to well-being, which relates to the sense of *the unaccountable* that is perhaps closest to oblivion: that which is obscure to scrutiny, resistant to rational explanation, mysterious, or ineffable. The unaccountable in this sense is that which resists the exhaustive explanation that we aspire to when we speak of accounting for some phenomenon or action. Although as agents we need to be accountable for ourselves, that capacity is imperfect. The limits of self-knowledge, self-control, and self-determination mean that we can never fully account for ourselves. There is no place to stand outside of oneself or one's life from which to conduct such an accounting, and in any case we all have our blind spots, and dark spots that refuse to surrender their opacity no matter how long we stare into them. But these lacunae are not—or not only—failures of introspection and self-governance; they are also sources of the sense that one's life has meaning, depth, and a quality essentially resistant to instrumentalization and control. Poetry and music gesture in the direction of these ineffable zones of human experience; tragedy gives them moral character. That such unaccountable elements of the self exist is an empirical claim, but we can make it normative and turn it toward privacy by adding the thought that this zone of experience and self-relation is unaccountable not because it is hidden or secret, but rather because something is lost in the attempt to put it on display or translate it into information.[33]

The rest of this book will be dedicated to defending these claims. So for now, let us linger on this last aspect a little more, and strengthen its connection to the discussion of possibility above, by using it to distinguish once again between the oft-elided concepts of privacy and secrecy. If what Foos did to Ben in *Voyeur* was perceive some secret of his, which he never subsequently communicated, then we would say that Foos was now in on the secret, that the pair of them shared it, and not that he deprived Ben of his secret or destroyed his secrecy.[34] By contrast, Foos cannot share Ben's privacy, only destroy it. This third aspect of being unaccountable gives an explanation as to why. For something to be a secret, you have to know what it is; it cannot be obscure, ambivalent, or inchoate. For something to be unaccountable in the sense I have been developing here, it must be just the opposite. The unaccountable is lost or destroyed when it is capable of being known in the propositional way that secrets are, as one thing or another.

Mapping the Unaccountable

Although our discussion of the unaccountable will be confined here, where it has proven useful for understanding the importance of privacy to agency, we should nevertheless pause a moment to consider a possible objection to this language. The objection I have in mind is the idea that privacy is somehow connected to *moral* unaccountability, in addition to the epistemic sort I discuss here, which will help bring out two of the more common arguments leveled against privacy—one good and one not so good.

The good argument comes from a longstanding feminist concern with the idea of a private or domestic sphere whose border marks the normative limit of justified state or social interference with the life of individuals and especially with the goings-on of the traditional nuclear family.[35] In its most common and powerful form, this argument is not actually a critique of privacy per se but of the public-private dichotomy of liberal political theory and politics, which arose in response to the fact that, not all that long ago, the nuclear family was treated by the law and civil society as a domain that was at once ruled by men and beyond the justified reach of state and social intervention. Throughout the history of this patriarchal regime—which, of course, was itself a confluence of public and private power operating under cover of the strict separation between the two—women lived a kind of semi-feudal existence, in which they were not only denied the same rights as their husbands but suffered domestic abuse and more that was, both as a matter of law and of culture, thought to be nobody's business but the family's. Rape-shield laws and "rules of thumb" are notorious examples, but harms to women and girls were pervasive. Men who raped, battered, or otherwise dominated the women and girls in their households were not held morally or legally accountable for their actions. This is obviously terrible and reprehensible, and it is in part to avoid any implication that I am arguing for this sort of moral or legal unaccountability that I have placed so much stress on the epistemic nature of my concept.

But perhaps one finds this distinction between the moral and the epistemic spurious. Indeed, there is another, somewhat less common strain of argument that sees the obscurity of privacy itself as providing cover for domination. The idea is that even if the problem of the liberal public-private dichotomy were solved, barriers to perception would still serve as cover for domination and other forms of wrongdoing.[36] It would be difficult to disagree with this objection to the *uses* to which privacy is put. (It's also an objection to the uses to which freedom or autonomy are put.) However, the response here is not to do away with privacy (or freedom), but to make it compatible with egalitarian, pluralist, and feminist values. In Seyla Benhabib's words, "After two decades of criticizing the private/public split, and the way in which this dichotomy has served to camouflage domestic violence, child molestation, and marital rape in the private realm, contemporary feminist theory is entering a new phase of thinking about these issues. The binarity of the public and the private spheres must be reconstructed, and not merely rejected."[37]

We should follow Benhabib and say that the goods of privacy are like other fundamental goods in another way: they are subject to questions of distributive justice. It was in this vein that second-wave feminists critiqued the domestic sphere of midcentury patriarchy as both insufficiently public *and* insufficiently private for the women relegated there.[38] More recently Simone Browne has argued that we ought to consider privacy (indeed, unaccountability) among the goods that a racist society unjustly distributes to the detriment of marginalized racial groups.[39] Édouard Glissant provides the rallying cry: "We clamor for the right to opacity for everyone."[40] Insufficient access to privacy or privacy in the fullest sense is a failure of justice with which everyone should be concerned.

However, to the extent that this critique of privacy is not a subset of a more general argument against abuses and maldistributions of all sorts of goods but regards the epistemic barriers of privacy as having some type of specific connection to wrongdoing, it is a species of an even more common argument: that privacy allows bad people to do bad things. This is not a good-faith argument but a piece of propaganda. For one thing, it is unusual that privacy would be singled out as permitting bad actors to do bad things, since other enabling conditions (again, like liberty, agency, and the rest) are never attacked in this way. It also attacks a strawman: nobody thinks that privacy is *for* doing bad things or getting away with them, but rather that it is for something else connected to human well-being. The concealment of privacy, like secrets or solitude, can be used for a variety of ends, some of which naturally might be described as abuses. It is against such abuses of privacy that this argument should be aimed, not privacy itself. But it is easy to argue against the abuse of *anything*. Rather, what this argument needs to show is that privacy belongs to the class of things that are so dangerous that they must be heavily regulated or abolished, and not to the class of things—like liberty, agency, intimacy, and so on—that we regard as vital to human flourishing and discredit only in their abuse. And this it cannot do.

We could go on. For instance, we could notice that once we move beyond the consensus examples of wrongdoing like domestic abuse, the sorts of "bad acts" obscured by privacy begin to differ according to one's politics. Is privacy bad if it obscures a girl's search for a safe abortion in a state that prohibits it? If it lets one burn a flag or worship Satan? It probably depends on who you ask. You can see the danger. Anyway, the danger also indicates that at another level of analysis, this sort of rational argument is beside the point. The claim that privacy promotes or lets people get away with bad acts isn't part of a good faith argument about the value of privacy, but rather a piece of propaganda meant to undermine its value, that has less to do with the adjustment of human circumstances to a picture of well-being than it does with the aims of surveillance and domination.

Unaccountable Intimacies

If we put ourselves in Ben's shoes, we might want to object that when we are in private we do not always or necessarily inhabit a state of unaccountability. True enough. Just as it would be exhausting to persist ceaselessly in being ourselves, it would be terrible to be always manifold or ambiguous, to have no direction or sense of oneself as valuing one alternative over another, and to never have anyone take seriously the relation between oneself and one's appearance. Just as the possibility of reintegrating into a community is necessary for the experience of solitude to be restorative—otherwise it becomes exile and alienation—the value of being unaccountable depends on its dialectical relation with accountability. The interests I have discussed in this chapter are interests in the healthy functioning of agency and human well-being more broadly, and not in the avoidance of recognition or the consequences of one's action as such. It is important to notice here that the relationship goes both ways. As agents, we also have an interest in being taken seriously by others, which includes their taking seriously the idea that we are as we appear to be. These interests contribute to the same picture of unalienated agency and broader well-being to which unaccountability belongs, as well.

It is beyond the scope of this discussion to give the full account of human flourishing in which being unaccountable would feature. My aim here is simply to have begun to make it plausible that such a picture exists, and that it provides sufficient, if not irrefutable, support for my explanation of the ordinary view of *Voyeur*. The truth is that I find this view of privacy and well-being not just plausible, but correct and compelling, in part because it directs our attention to areas of value overlooked in debates about privacy and underappreciated in contemporary life. We will have much more to say about the broader view of oblivion and well-being later in this book, so for now let us draw the discussion toward a close by connecting it to something else that makes a life go well: physical and emotional intimacy, whose relation to privacy becomes clearer when seen through the lens of the unaccountable.

The view I have presented of how Ben is worse off simply by being perceived involved an account of a moral right to privacy based not in interests in control, autonomy, or self-determination, but just the opposite. This is, as we have seen, at odds with a great deal of thinking about privacy today, which understands privacy's value as a function of the control it gives individuals over their personal information or the perceptual access that others may have to their persons and things. One prominent version of this idea is that such control is necessary for creating and maintaining intimate relations, on account of how privacy permits us to share certain things with intimates that we do not share with the broader range of our social relations. James Rachels offers one of the more interesting and influential versions of this idea. Privacy is valuable, he thinks, because it is necessary for maintaining a variety of social relationships, and it is necessary to that end because different relationships are characterized by different levels of what is appropriate for others to know about us. We don't want our parents to know all that our lovers know about us and vice versa; certain information may be appropriate to share with friends but not students. Rachels supports his argument with an example drawn from a marvelously unexpected source: the TV-guide section of the *Miami Herald*.

> "I think it was one of the most awkward scenes I've ever done," said actress Brenda Benet after doing a romantic scene with her husband, Bill Bixby, in his new NBC TV series, *The Magician*.
>
> "It was even hard to kiss him," she continued. "It's the same old mouth, but it was terrible. I was so abnormally shy; I guess because I don't think it's anybody's business. The scene would have been easier had I done it with a total stranger because that would be real acting. With Bill, it was like being on exhibition."

Here is what Rachels thinks this story reveals about privacy:

> I should stress that, on the view that I am defending, it is not "abnormal shyness" or shyness of any type that is behind such feelings. Rather, it is a sense of what is appropriate with and around people with whom one has various sorts of personal relationships. Kissing another actor in front of the camera crew, the director, and so on, is one thing; but kissing one's husband in front of all

these people is quite another thing. What made Ms. Benet's position confusing was that her husband was another actor, and the behavior that was permitted by the one relationship was discouraged by the other.[41]

Let us put aside the idea that the spousal relationship somehow discourages kissing in front of a film crew, which seems clearly wrong to me. On the most obvious reading, Brenda Benet's statement of discomfort does not support Rachels's theory. Surely the cast and crew would not be surprised or scandalized to learn that a husband and wife share kisses; surely it is not inappropriate for them to have such knowledge. Anyone would have already assumed as much (and more) without having to give it a thought. It is hard to see how *any* of Brenda's relationships would be damaged by this. The crew learns nothing by witnessing a husband and wife share a smooch, unless, in a very literal and technical sense, they gain something along the lines of "the knowledge that Brenda and Bill have kissed at least once." Of course no one thinks like this. Yet even if they did, it is very hard to imagine how that knowledge would damage or even alter Brenda or Bill's relationship with the crew or each other. The contorted expression of what the crew could possibly have learned information-wise about Brenda and Bill from their scene is yet another indication of just how odd it can be sometimes to think of privacy in terms of controlling "information flows." Again, we encounter the distinction between privacy and secrecy: Brenda may not want anyone to see her kiss the actor with whom she's having an affair because she does not want that information to get out, but of course that isn't the case when it comes to kissing her husband, Bill. Instead, Brenda's statement "I don't think it's anybody's business" echoes the typical response to the invasive question: it doesn't invoke a barrier behind which information is kept from some but not others, but one that protects against the creation of information or knowledge in the first place.

Rachels is right to think that matters of privacy, information, and intimacy are involved here, just not in the way he supposes. What discomfits Brenda is the sense of "being on exhibition." By this she cannot mean that what is exhibited is some previously unknown piece of information about her private life, in part because of ordinary assumptions about the love lives of married couples, but also because the romantic scene does not in fact give *any* information about what Brenda and Bill are like beyond the limits of the scene. Because the romantic act is a scripted piece of explicit make-believe, it gives the crew little reason to think that what they see reflects in any way on how Brenda and Bill are beyond the bounds of "action" and "cut," although it may say something about the quality of their acting. Nor does the complaint about exhibition have to do with the sort of behavior on display being the kind of thing that is properly hidden—she admits that she would not have minded going at it with an actor who was not her husband. Our discussion of unaccountable experiences helps to make sense of Brenda Benet's complaint by illuminating the opposition between the awareness of being on display, on one hand, and real physical intimacy, on the other. "Being on exhibition" is contrary to intimacy with her husband, but not the play-act of intimacy conducted with a stranger, because it is deleterious of the psychological unaccountability that typically attends the fullest experiences of physical intimacy.

As in Ben's case, Brenda's exhibition is detrimental not because of norms governing what is appropriate for others to know, but because being on display engenders the kind of self-consciousness that is at odds with being unaccountable. When it comes to kissing an actor who is not her husband and being "on exhibition" while she does it, Brenda doesn't care about how exposure and self-awareness undermine the unaccountability of physical intimacy because she does not want to have that sort of experience in the first place. There is risk in the real experience of intimacy that is absent from the pantomime: the disappearance of the accounting self makes one vulnerable, open to the unstructured and uncontrollable (animal, even) parts of oneself. Much like the statesman and the divine in the previous chapter, "down upon all-fours," in experiences of physical intimacy one gives oneself over to unpredictable potentiality at the core of the person and encounters its force within oneself—and within the other, as well, which is why the experience can be so exciting but also frightening.

The most minimal acquaintance with physical intimacy is enough for one to know that it is an experience in which the self-conscious agent tends to disappear. The self as well as the boundary that separates it from the other seem to dissolve while conscious attention and embodied presence become, if anything, heightened and more intense. This experience shares certain key features with Taylor's state of radical openness in which we encounter the deep, inchoate, and unstructured dimensions of the self, except that in the case of physical intimacy, we encounter it in the other at the same time. In the act, individuals temporarily relinquish deliberate, rational control of their lives and bodies in order to encounter the regions of the self and its relations that lie beyond knowledge and control. This is why a failure to lose oneself in the act is commonly thought to be an error of lovemaking, to which the natural response is, once again, "you need to get out of your head" or "let yourself go a little."

It is on account of limitations of knowledge and control that we consider sex one of the deepest, most intimate ways of being with and coming to know another person.[42] But it is not just others that we come to know in this way. We also learn about ourselves. Experiences of physical intimacy can be occasions for self-discovery in addition to pleasure. However, since the object of discovery, aside from predilections and pleasures, are the unaccountable regions of the self, these discoveries of self-knowledge inevitably seem to lose their force when we try to put them into words. Even a good kiss is like this. Pablo Neruda writes, "Love I learned in a single kiss/and could teach no one anything," capturing at once both the revelatory power and deep incommunicability of the kind of understanding one gains from such an experience.[43] A description of sex, or even a kiss, is pornographic by comparison to the experience itself because even the sharpest, most exhaustive expression seems to miss the point or to reduce it to cliché, schema.[44] The experience is deflated by description, but also by the descriptive form of self-awareness that consists in the real-time accounting for oneself via narration, justification, or the consideration of how one is being interpreted by others. Against the many exigencies of accounting for oneself, privacy's protection of space and time for the unaccountable offers us the chance to live with our murky parts by being murky ourselves, to experience the bits of ourselves and others that lie beyond our control and understanding. No wonder we seek privacy to enjoy the fullest experience of losing ourselves with others.

3

HIDING IN PRIVATE

For years I commuted from Providence, Rhode Island to Cambridge, Massachusetts, by train. One morning, I boarded the train to find the typically quiet cars of sleepy office workers packed with revelers decked out in Red Sox gear, already half-drunk and shouting at eight o'clock. All the carriages were standing room only, and I found myself sandwiched between two groups of rowdy fans who, I soon learned, were on their way to the World Series victory parade in Boston. For a while I was among the crowd, talking with strangers, overhearing conversations, politely declining the many generous offers of schnapps and cinnamon-flavored whiskey from my fellow travelers. But then at some point, I pulled out my phone and began to send text messages about the situation to my wife, my brother, and a few friends. Their replies arrived almost instantly, and then, as strange to say as it is common to do, I found myself engaged in several conversations at once. It was as if I was suddenly absent from the train. The ruckus fell away. I stopped feeling the bodies of others brushing up against me; my own body even seemed to recede from physical presence as my mind focused on the messages I sent and received in rapid-fire succession. I was, as we say, elsewhere, because my mind was elsewhere. My mind and attention were actively engaged with parts of the world other than the one my physical being inhabited.

The experience recurs frequently in less exotic circumstances—indeed, anywhere I carry a smartphone. My wife and I are together, in private, and one of us looks down at our phone, our attention recalled to the device either by the ping of a notification or the Pavlovian habits we've developed over the years. When I look down at the screen, she knows immediately that I am no longer with her, not fully, but that I am at the same time also attending to someone else or gazing through a little handheld window onto other places and other lives. When I am home alone, in private, I am hardly ever really alone anymore. Am I still in private?

My experiences of dislocation, fragmentation, and attention may seem quaint in our era of constant connectivity. By *connectivity* I mean to refer broadly to the mobile communications technologies that are, in Sherry Turkle's phrase, "always-on/always-on-you" and the complex of norms, expectations, behaviors, and habits they enable and promote.[1] Anyone who owns a smartphone is constantly engaged in a dance of attention and presence far more complex than the sending of text messages from a crowded train. We conduct multiple simultaneous conversations with friends and family, scroll rapidly though videos shot all over the world, post our thoughts for a globally dispersed audience, and swipe through profiles of potential romantic partners, all more or less at once and with the thoughtless ease of second nature. Although this great power of connection to other lives and worlds was initially welcomed with enthusiasm, soon a darker note began to sound. "As our mental lives become more fragmented," writes Matthew Crawford, "what is at stake often seems to be nothing less than the question of whether one can maintain a coherent self."[2] Bernard Harcourt also thinks that recent changes in our habits of attention threaten no less than "the mortification of the self."[3] Justin E. H. Smith calls it a "crime against humanity."[4] The consensus about the dangers of new technologies is almost as widespread as their adoption: Our constant connection to others and virtual publics in the digital age is commonly thought to pose a serious threat to well-being and a diminution in the quality of our privacy.

Yet, the idea that new media poses a threat to subjectivity because of its effects on our attention was of central importance to the earliest advocates for a right to privacy in the nineteenth century. For instance, an 1874 column in *The Hartford Courant* titled "Is There Any Privacy?" complains of how the new media of the day "invades the study of the clergyman, the office and house of the politician, the sanctum of the professor, and demands to know of each what he thinks about the new book or the latest scandal, and whether he drinks wine or milk with dinner, and what he thinks of his mother in law."[5] You would be forgiven for thinking that the columnist was talking about social media and not, ironically enough, newspapers. John Ruskin restates the problem to make clear that it could not be laid exclusively at the feet of the new media fueling what contemporaries described as "the age of publicity"—a phrase that today seems as short-sighted as "The Great War" while also indicating, once again, a continuity of concern with our own time. The individual, thought Ruskin, also bore some responsibility for inviting too much publicity into the private sphere, with the result that the privacy of the home was being transformed into something worse and less sustaining. "The true nature of the home," he writes,

> is the place of peace: the shelter, not only from all injury, but from all terror, doubt, and division. In so far as it is not this, it is not home; so far as the anxieties of the outer life penetrate into it, and the inconsistently-minded, unloved, or hostile society of the outer world is allowed by either husband or wife to cross the threshold it ceases to be a home; it is then only a part of the outer world which you have roofed over and lighted a fire in.[6]

Ruskin's view of the dangers of penetrating publicity was widely shared among his contemporaries. "We have reached an age of publicity," proclaimed George Lorimer in an 1892 sermon later reprinted in the *Boston Globe*. "And the result of it is that we forget ourselves."[7] Henry James's journal entry for November 17, 1887, reads, "One sketches one's age but imperfectly if one doesn't touch on that particular matter: the invasion, the impudence and shamelessness, of the newspaper and the interviewer, the devouring *publicity* of life, the extinction of all sense between public and private."[8] Warren and Brandeis sought to express and solidify the common thinking on this aspect of privacy, too, arguing that the advent of mass-market newspapers harmed both individuals and society by offering a constant supply of novel and exciting but essentially frivolous information: "Triviality destroys at once robustness of thought and delicacy of feeling. No enthusiasm can flourish, no generous impulse can survive under its blighting influence."[9] Although their concern sounds in a decidedly more Romantic key than we find in James, Harcourt, and Crawford, the understanding of the threats to well-being posed by new information technologies and the habits of attention they foster is essentially the same.

The idea that one's privacy is diminished—invaded, even!—by the publicity that one invites into one's life, and by publicity of the wrong sort, was a constant refrain throughout the twentieth century, as well. For instance,

Wallace Stevens worried about the increasing presence of radios in homes, lamenting that the imagination of his time suffered from being overconnected to the world at large and therefore stunted by the "extraordinary pressure of the news" streaming into more and more homes over the airwaves. "We are close together in every way," he writes. "We lie in bed and listen to a broadcast from Cairo, and so on. There is no distance. We are intimate with people we have never seen and, unhappily, they are intimate with us."[10] In 1946, Gertrude Stein echoed Lorimer's sermon half a century hence: "Everybody gets so much information all day long that they lose their common sense."[11] The following decades saw similar complaints about rising popularity of television.

> What now predominates through TV at home is the—real or fictious— broadcast outer world: and this reigns so absolutely, that it thereby makes the reality of the home—not that of the four walls and the furniture, but communal life—phantomlike and invalid. When the phantom becomes reality, reality becomes a phantom.[12]

Today we have largely abandoned the idea that one's privacy can somehow be diminished by how one spends one's attention, in addition to questions of snooping, surveillance, and exhibitionism. The reason for this, as I suggested in the introduction, is that the previous understanding of privacy was gradually supplanted by a neoliberal, voluntaristic conception according to which privacy is more like personal property or a preference rather than a fundamental interest.[13] If the value of privacy depends on our choosing it, then the normative question of somehow diminishing it by inviting a certain quantity or quality of publicity into our lives is rendered moot by the simple fact of our having consented to it by the invitation. Yet there is wisdom in the abandoned view that we would do well to rescue.

In this chapter I will make the case for recovering the older ideas about attention and privacy, or at least parts of them, which in turn will further develop my argument about the importance of privacy for well-being. But I will proceed by what might at first seem a somewhat circuitous route, beginning not by talking explicitly about attention but about the difference between hiding and privacy. This difference—which, as we will see, holds moral significance—largely turns on the question of what one's attention is like. This has serious consequences for the standard theoretical and political views of privacy, which have a hard time expressing the difference between hiding and privacy and therefore tend to miss some of that moral significance. If we lose sight of the difference between privacy and hiding, as it seems we largely have, we run the risk of basing our defenses of privacy on an impoverished concept.

A particularly glaring example of such hamstrung moral reasoning is the idea that privacy is of value only for those who have something to hide. Obviously, this slogan is ill-intentioned. It has been used for centuries to pressure citizens, co-religionists, and family members to render their lives more transparent to the mechanisms of power; more recently, it has served as a potent justification for the expansion of the surveillance state. What is most maddening about the slogan and the responses to it is that no one seems to notice that it is also specious. *Hiding* is for those who have something to hide. Privacy, we ought to say, is for something else. To say this, however, we will need to unlearn the elision between hiding and privacy and to understand what makes the two forms of concealment distinct. For if hiding and privacy are just two ways of saying the same thing, then we have reason to think that people with nothing to hide maybe really do have nothing to fear from surveillance. At most they lose the ability to hide, which invites the question of why they might want that ability in the first place. This, in turn, feeds the hermeneutics of suspicion one finds in the surveillance state and beyond that views simple concealment as giving reason for suspicion and warrant for surveillance. However, if ordinary usage is correct and the two concepts are distinct, then even those with nothing to hide have something to lose: namely, whatever it is that privacy gives them. Getting clear about what we stand to lose in such a case, and why we should care, is surely one of the more pressing political and ethical questions of our time.

Another reason to look to hiding is that in our era of mobile connectivity and social media, the condition of our privacy has begun to look quite a bit more like hiding than it used to. That is how it appears to me, at least, and to a great many others whose views are reflected in ethnographic studies of the connected life. To be sure, it is a bit paradoxical to think that we could experience a state like hiding thanks to our constant connection to one another and discursive publics. Likewise, the culture of exhibition, confession, and "oversharing" promoted by social media certainly appears more like a cacophony of calls for recognition than the intentional avoidance of discovery.

Nevertheless, there are certain phenomenological and normative similarities between hiding and personal seclusion in the era of constant connectivity. Most notable among these is that they are both forms of experience in which being cut off from the common world depends upon a type of connection to it; both experiences share certain psychological characteristics; and both, all things considered, tend to be normatively disfavored, as opposed to, say, the broadly desired seclusion of privacy or the tether of being in love.

Finally, I want to connect the question of privacy, attention, and connectivity to another problem of the digital age that is also thought to be at once novel and pressing. This is the irony that although we are more connected than ever, we are at the same time more isolated, lonely, and alienated. The link between connectivity and alienation will emerge from our distinction between hiding and privacy, so for now allow me to plant a seed in your mind. Compare Stevens's complaint about the radio to Hannah Arendt's account of how a change in the quality of human seclusion from solitude to loneliness lays the ground for, and is in fact a means and goal of, totalitarianism: "It fits [a person] into the iron band of terror even when he is alone, and totalitarian domination tries never to leave him alone except in the extreme situation of solitary confinement. By destroying all space between men and pressing men against each other, even the productive potentialities of isolation are annihilated."[14] The iron band of terror that is felt even when alone is another way of describing the fear (or thrill) of a hider that connects her to the seeker, albeit with the psychological strain and stakes of discovery pushed to an extreme. It is a form of connection and psychological orientation to another that obtains even when in isolation. But "never alone" is also the basic condition of connectivity, and the degradation of solitude into loneliness is, as we shall see, analogous to the degradation of privacy to hiding.

Of course, I will have to say a bit more about hiding before I can make this clear and hopefully convincing, so for now allow me to offer an image and a preview of where I'm going. The image is of two eyes. The first is the strained eye of the hider staring out of her hiding place, looking for those who are looking for her; the second is the strained eye that stares into the smartphone or computer screen. Two exhausted eyes persisting in the activity that exhausts them. The analogy of these eyes reflects certain similarities in the sort of experience that lies behind and is belied by them. Now for the preview. Privacy and hiding are not the same, and it is bad to confuse them. Our current experience of what heretofore had been privacy looks quite a bit like hiding (that is, our eyes fixated on virtual publics really are like the eyes of the hider looking out in strained expectation for the appearance of the seeker), which is bad for a variety of reasons, all of which a theory of privacy ought to address.

In Hiding or in Private?

What is the relation of hiding to privacy? Think of a child who, engaged in a game of hide-and-seek, runs off and crawls into a hollow log. Picture a fugitive in a hayloft. Remember the Franks in their secret annex on the Prinsengracht. The natural way to describe what is going on here, as the name of the game would indicate, is that these people are hiding. When asked what the child, the fugitive, or the Franks are up to, no one would describe their concealment in terms of privacy. In fact, to describe the Franks in the language of privacy is not just inaccurate but perverse, notwithstanding the many ways their accommodations were indistinguishable from a private home. The mode of resemblance between the Franks' quarters and a private residence is that of travesty, for in important and perhaps obvious ways their secret annex was more like a hollow tree than the private home it could have been under other circumstances. This sense of travesty, along with the fact that in English we naturally refer to their domestic space as "the secret annex" and "hiding place"—not "private annex" or "private space"—indicates that there is a meaningful difference here. The feeling of perversity suggests that this difference holds moral weight.

Is it good or desirable to be in hiding? Is it an important part of a human life, an aspect of its going well? The answer seems to be, "It depends." On the whole, hiding holds for us a negative connotation, although it is also something we would like to be able to do if necessary. In a good life, hiding's value is exigent and derivative. It can form part of exciting games for children and lovers, but ordinarily nobody would choose to hide absent some external compulsion. Ask the same question about privacy, however, and get the opposite reaction. For however one defines it, privacy is typically thought to bear an important and fundamental relation to well-being, the sort of thing we might choose for its own sake.

So far, there is nothing particularly unusual about this verbal distinction. Every language is full of near synonyms highlighting salient differences between otherwise similar objects. Nudge, push, shove. Hidden, secret, private.

The matter only appears strange or counterintuitive when approached from the perspective of privacy, especially theories of privacy. For the standard theories of privacy fail to distinguish between hiding and privacy or to account for the sense of perversity and travesty at calling the Franks' secret annex their "private annex." This too might seem like a problem only for philosophers and nitpickers. Yet, it turns out that the challenge of distinguishing hiding from privacy is bound up with, and will help us to understand, several pressing challenges of the digital age.[15]

Let us begin with an uncontroversial example of privacy. Picture a girl alone in her room with the doors closed and curtains drawn, and assume for the moment that the year is 1999, or 1899, and that she has no access to a smartphone or Internet connection. Notwithstanding disagreements about what privacy is, no one disagrees that this girl has it. Among theorists of privacy today, the standard way of understanding her privacy would typically draw upon notions of "control" and "access."[16] According to the standard views, we are correct in saying that this girl has privacy either because she has control over who can know a certain piece of information about her (that is, whatever you might learn if she opened the door to you) or because there are "extraordinary limitations" on others' ability to access her or "information" about her. But this is also true if we move the girl from her bedroom to the log; it is as true about the Franks on July 6, 1942, in their secret annex as it was a day (or decade) before, behind the closed doors of their own home.

Let us examine some of the competing attempts to describe what we mean when we say that the girl has privacy. Because I draw the following accounts from the field of analytic philosophy, permit me a word or two to prepare readers unfamiliar with the discipline for what might seem like unnecessary abstraction or the type of abstruse hair-splitting incentivized by the economic structure of the academy. There is probably no place in all the world, certainly not in the English-speaking part of it, where the question of how to understand privacy is more hotly debated than in the field of analytic moral philosophy. These debates engage many brilliant minds around a host of creative examples and approaches, though they also often employ a highly abstracted formal language intended to pinpoint as precisely as possible just what we talk about when we talk about privacy. In the following examples drawn from that debate, please ignore the pleonasm "*informational* privacy," which, as we have already seen, is a vestige of the misguided distinction between the ordinary meaning of privacy and so-called decisional privacy. What is important to keep in mind is that each of these rival accounts seeks to explain what we mean when we say that the girl in her room has privacy, and that taken together, they represent the lion's share of contemporary thought on what privacy is and how it structures our relations with others.

> "An individual A has informational privacy relative to another individual B and to a personal fact f about A if and only if A controls whether B knows f."

> "An individual A has informational privacy relative to another individual B and to a personal fact f about A if and only if there are extraordinary limitations on B's ability to know f."

> "An individual A has informational privacy relative to another individual B and to a personal fact f about A if and only if (1) f is undocumented and (2) B does not know f."

> "An individual A has informational privacy relative to another individual B and to a personal fact f about A if and only if B does not know f."

> "An individual A has informational privacy relative to another individual B and to a personal fact f about A if and only if B does not actually access f."[17]

The stripped-down analytical language in these definitions is supposed to capture privacy in its essentials, yet if they cannot distinguish privacy from hiding—or from secrecy, for that matter—they can hardly be said to have achieved their object or even completed the analysis. If privacy only exists to prevent access to some piece of information (or one's person), then it would seem that the girl in the log also has it. Likewise the Franks in the secret annex. And it won't do to say that the hider may be found any minute by the seeker, or that the one who is hiding is vulnerable in a way that one is not when in private. A hider might not be found—good ones often aren't— and in any case the same is true about privacy. That one has privacy does not mean that a snooper will not come around and pierce it. Privacy can be undermined, broken, and destroyed just as much as hiding. The same is true if we try to reframe the reply in the mode of republican political theory and say that, until found, the hider lives under constant threat of discovery. The one in private is also under constant threat of having their privacy violated, it's just that the stakes tend to be somewhat lower (although not necessarily).

These theories fail to accurately describe privacy because they wish to say what it is without considering what it is for. Theories of privacy are supposed to come in two varieties: "descriptive" accounts, which purport to analyze and describe only what we mean by our concepts and practices of privacy, and "normative" accounts, which consider its moral basis and role in human well-being. The former is thought to have the advantage of not relying on prior moral or political commitments and therefore to be more objective and relatively insulated against normative objections. Yet as the failure of the above theories even to *describe* what makes privacy different from hiding reveals, this is another of the false distinctions that muddle our understanding of the subject. When we talk about privacy, we are talking about a complex of normative social practices, and practices are always *for* something beyond or in addition to even the most exhaustive account of how they work. A description, even a *mere* one, that avoids the end of a practice will be so superficial as to hardly merit the name.[18] You simply cannot understand the difference between the two girls, or the Franks' two apartments, or a hidden life and a private one, without broaching the question of what privacy is for.

The failure of so-called descriptive accounts to capture the difference between hiding and privacy shows that if you want to understand what privacy is, you must know what it is for, and that whatever privacy is for is somehow inimical to what hiding is for. The reason for thinking this is that as forms of concealment, privacy and hiding would appear identical but for the ends to which they are put and what they are like to experience. This means that what privacy protects, like hiding, is something more than simply a state of affairs with regard to information.[19]

At this point, an objection might be raised. What if hiding is merely a subtype of privacy? That is, what if privacy concerns the general concept of barriers to unwanted perception, while hiding picks out one type of reason for wanting those barriers? But it is not true that privacy is necessary for hiding. You can hide by going incognito or adopting a false identity in public; you can hide from the law by being "on the run"; and you can hide in plain sight like Poe's purloined letter or by mixing into a crowd. Anyway, if hiding were a subspecies of privacy, it would only exacerbate the puzzle of the perversity in describing the Franks' secret annex as their private one. Instead, we would be better off describing both hiding and privacy as distinct subtypes of concealment. Since our goal is to get clearer about what is unique about privacy—that is, what, as a mode of concealment, distinguishes it from hiding—we cannot start by giving a definition of privacy. Instead, let us begin by thinking about the far less contested concept of hiding as a way of describing privacy by reference to what it is not.

Abjection of the Hidden

The most obvious fact about hiding is that it depends upon the existence of a seeker. The glaring difference between the Franks in their prewar apartment and later in their secret annex is that in the latter their concealment is a response to somebody trying to find them. The motivation behind and the value of their concealment is connected to a particular end: the prevention of discovery by those who would find them out. The same is true about the child in the log. One can hide oneself, an object, or some information from a single individual, a group of them, or all the world, but hiding is essentially hiding *from* someone or something. We hide from the past, from our responsibilities, and so on, which, significantly, are always "catching up with us." The thing from which one hides need not be harmful or even undesired—the girl in the log does not lose the game by being discovered, and we often hide ourselves or items with the express intent that they may be found—but in every case the seeker is the ground of what is hidden. This is true even in cases of paranoid delusion, where someone feels that they are sought by others when in fact they are not. Although the seeker is imagined, it is nonetheless fundamental to the condition of hiding.

This dependence of hiding upon the seeker has consequences for how we understand and value what is hidden. The binary is not evenly matched: the seeker can dissolve the hider-seeker relationship *and know it*, but not the other way round. What is hidden can only endure, persist, wait; the hider can also give up hiding, but only at the cost of losing whatever they had hoped to achieve. (The difference between coming out of hiding and coming out of privacy is also illuminating here.) This is not to say that hiding is necessarily passive, although even its most active forms, like being on the lam, consist largely in endurance and waiting and in any case are centered on discovery, if only by constant attention to its avoidance. The position of the hider is relatively abject by default, which is why the excitement of hide-and-seek is unlike that produced by other games. Whereas most games are exciting because of competition and the chance of winning, the thrill of hide-and-seek is having not yet lost. The game, like the broader experience of hiding it distills, is characterized by that "not yet" or "still alive, still hidden," which is why a lazy seeker can ruin the game for everyone while bad hiders do not.

The possibility of discovery is a necessary condition of something or someone being hidden—otherwise what is hidden becomes lost. I think this is what motivates the assumption, often latent but always active in our thinking about concealment, that discovery is not only the essence of hiding but its inevitable conclusion. This is surely one of our oldest ideas. St. Luke: "Nothing is secret, that shall not be made manifest; neither any thing hid, that shall not be known and come abroad."[20] Shakespeare: "Truth will come to light; murder cannot be hid long; a man's son may, but at the length truth will out."[21] Freud: "He that has eyes to see and ears to hear may convince himself that no mortal can keep a secret. If his lips are silent, he chatters with his fingertips; betrayal oozes out of him at every pore."[22] The presumption of discovery seems to support the additional inference that what is hidden is ipso facto contrary to the hider's self-interest, for if what is hidden will eventually be revealed, what other reason could there be to forestall discovery? The inference is faulty but nevertheless exercises a potent influence over the moral psychology of hiding. Imagine, for instance, what your first thought would be if you opened the door to my office and saw me quickly conceal something behind my back. Even if we ought to do away with this bias, we need to be aware of its connection to our concepts and practices of hiding so that we may notice and resist the drift or deliberate insertion of these associations into our understanding of privacy.[23]

The contrast here between hiding and privacy could not be clearer: privacy is not dependent, conceptually or ontologically, on the possibility of being discovered. We do not need to know anything about who would discover the girl in her bedroom to describe her as being in private, only that there are certain barriers to her being perceived and, of course, that she is not hiding. It's simply not relevant whether anyone is or ever will be interested in invading her privacy for her to have it or to want it. Unlike the paranoiac, she is neither irrational nor mistaken if she seeks privacy when in fact no one is trying to find or discover her, although we might say that to the extent that she does seek privacy as a way of avoiding detection from particular others it begins to resemble hiding, even if it does not tip over all the way from one condition to the other. This observation gives additional support to a claim I made above when discussing the so-called descriptive theories of privacy: that the value of privacy must have to do with more than the mere keeping of oneself, an object, or information away from others, be they a single seeker or the world at large. Privacy is not, in other words, for keeping secrets or hiding things, although it might also achieve that end, in the same way that a frying pan may be used to shovel snow.

Tethered Experience

Hiding is also a form of experience and activity with its own particular intentional and psychological characteristics. You cannot hide something by accident, for instance, only lose it. The question of whether Gretel is hiding in the forest or is lost in it will depend on her understanding of what she is up to.[24] It is in this sense that we can speak of one who is no longer being sought as hiding, or of how a delusional person hides beneath an imaginary shroud, but only if they are actively engaged in the activity of hiding. If the girl fell asleep in the log and was left behind, or if she returned the following day to get another look at the bugs who live there, she is no longer hiding, not really, even though she is nevertheless concealed in a way we might describe as being hidden.

The form of experience captured by the active form of the verb "to hide" and the adjectival "in hiding" (as distinct from the passive "to be hidden") is best understood not primarily as a conscious activity but something more like a psychological tendency, a frame or habit of mind, or a state of affairs with several conditions, not all of which need be present at once. To be sure, I cannot hide without some understanding of what I am doing, even if I may unwittingly cause myself or something to be hidden. But such awareness need not be constant. Anne Frank did not cease to hide when distracted by her correspondence courses or falling in love with Peter van Pels, but then suddenly was hiding again when the bleak facts of her situation returned to her. Yet if she were never aware that she was hiding—as, say, a much younger child may be fooled by her parents—then it would not be fitting to say that she is hiding in this active, experiential sense. She would merely be hidden. This is important because it means that we should understand the experiential aspect of hiding as something more like an inclination to think and behave a certain way or to have a certain form of experience. Such inclination may be personal idiosyncrasy or pathology, but more often it will come as a response to an individual's material circumstances. Therefore, judgment of whether someone's condition looks more like hiding than privacy (or something else) will not require knowledge of moment-to-moment psychological states but can be made at a higher level of generality and with reference to nonpsychological aspects of human life.

The phenomenology of hiding mirrors its conceptual and ontological dependence on the seeker. This is what accounts for the exquisite inner tension of hiding, in addition to the sense that the hidden inclines to its own discovery. The thoughts and attention of the hider reach out, as it were, to the seeker, while her body must remain concealed. Her mind rushes out to the seeker or at least fixates on her entanglement in the condition of being sought (again, as a general matter). Her thoughts, attention, and time are not entirely her own, but are oriented by her material conditions to the seeker. We might also say *social* conditions, which underlines the fact that hiding is a form of sociality. Whether one is hiding with many from one, or alone from many, it is a particular form of being involved with others while at the same time separate from them. In the game of hide-and-seek, this manifests as a feeling of excitement and anticipation; in the case of the Franks, oppression and terror. In the former, the footsteps of the seeker inspire a thrill; in the latter, "horrible fright" when they hear them, "constant dread" when they do not.[25]

It is important to notice that the quality of the fright and dread associated with hiding (or when sublimated through play, the thrill) is different from the fright or thrill of other scary experiences or games where, as the saying goes, one forgets oneself. Hiding, by contrast, is a powerful reminder of one's self-identity, if only for the simple reason that if one were someone else in the relevant sense, then one would not be sought. The hider is, to borrow from Stanley Cavell, impaled upon herself and at the same time cut off from the common world.[26] The hider is individuated in the sense of being "singled out," isolated by the seeker's attention. Individuality in this case—both one's non-fungibility as a discrete human being and the particular characteristics that make one identifiable as the seeker's target—becomes a sort of curse. This means that hiding is characterized not only by a particular relation to the seeker but also to oneself. To hide is to experience individuality at its most exquisite, to feel the sharp edge of self-identity. Perhaps children have an intuitive grasp of this, which is why in hide-and-seek they so rarely hide in groups. Hiding is a form of isolation grounded in the integrity of personal identity, albeit of a kind that does not typically fit into a picture of human flourishing. The contrast with our discussion of privacy in the previous chapter could not be clearer.

Consider the distinction between hiding and solitude, another form of isolation much less contested than privacy. Solitude is, at its most basic, "a time in which experience is disengaged from other people."[27] We seek solitude as a form of repose, "getting away from it all" in order to relax at a distance from the pressures of life among others, to hear the voice of nature, or simply to be alone with our thoughts. Hannah Arendt describes solitude as the conditions that permit one to split oneself in two, as it were, and get to know one's own mind through a kind of self-dialogue.[28] This is why the description of being "by oneself," rather than simply "being alone," seems to capture what has traditionally been thought to be valuable about solitude: it gives one the opportunity to get to know oneself apart from others and the distractions and pressures of life in society. Thus, solitude is undermined by an inability "to get away," an inability that is the very essence of hiding. Solitude provides the conditions to get to know oneself in the absence of the pressures of the social world (including those pressures that support) and therefore to learn to trust oneself, one's mind, and one's ability to get along with nothing but one's own resources. The experience of solitude might be pleasant in itself, or might not, but in any case it is thought to be valuable because of its relation to repose, authentic or independent self-knowledge, and personal self-reliance.[29] Hiding, it would seem, induces an opposite experience, which we will discuss in a moment.

But what conclusions can we already draw about the experiential dimensions of privacy by contrast to those of hiding? As a rule, theories of privacy tend not to posit positive psychological conditions of privacy, but only a certain state of affairs with regard to control over or access to one's person, information, and so on. Yet we have seen that a judgment of whether one is hiding or in private can depend on an individual's state of mind. If all we know about the girl in her room is that she is intent on not being discovered, we would say that she is hiding in there. And if we understand privacy by, among other things, its opposition to hiding, we can begin to draw some conclusions, for the experience of privacy must lack those qualities that distinguish hiding from other forms of concealment.

Whatever else it involves, the experience of privacy must lack the psychological orientation of being tethered to those "outside" of one's condition who would seek to destroy or pierce it. This is just to restate the nonidentity of hiding and privacy at a higher level of specificity. Recall the example of the delusional or mistaken hider, which showed that the experience of hiding can obtain in the absence of an actual seeker so long as the orientation is present. To the extent that one has this orientation, one is in a condition more like hiding than privacy, independent of a seeker's existence. Since we must understand the tether connecting hider and seeker as a general tendency or inclination to have this particular sort of attentional orientation to the outside world, we can conclude that one aspect of privacy is the exclusion of this tendency or frame as a general matter.

An experience of concealment that excludes this psychological tether will, among other things, not be abject in the way of hiding. This is not to say that one would be in control of one's attention, just that it is not directed outward in this way. Here the republican frame seems fitting: one is in private or has privacy to the extent that one's mind and attention are one's own, not in the sense of being controlled by some homuncular self enthroned behind one's mind and attention but simply not subject to forces beyond the bounds of one's privacy. Part of what it means to have privacy then, in the fullest sense, involves not being too fixated on the world beyond privacy's interior in the way that a hider fixates on a seeker. This is basically the privacy complaint made in the wake of each development in the nineteenth, twentieth, and twenty-first centuries that increased access to information and communication in the home, from newspapers and photographs to radio, television, and social media. We can find many more examples in today's technologies of mobile connectivity—both the digital publics designed to capture our attention and the tintinnabulating devices that recall our minds to them throughout the day.

It certainly is my experience that the constant connection to the digital public sphere, in which I can publish and participate at any hour of the day, has changed my habits of thought. I am less immersed in the book I am reading because my thoughts wander to what others are saying about it right now, what reviews and secondary literature I can access without getting out of my seat. I wonder, without meaning or wanting to, whether this paragraph or that one will make an engaging post on social media, whether this friend or that one would enjoy a picture of it. Group chats of friends and family far and wide constantly call for my participation in endless conversations. Although my body may be within the four walls of my little apartment, my mind, my attention—as much a part of me as my body—are elsewhere. Not in private, at least not in the fullest sense.

Definite and Indefinite Modes of Self-Relation

For a thing to be hidden, it must be definite. There must be some fact of the matter about it—it must be known by at least one person or documented, otherwise what is hidden becomes lost. Even when we say that something is "shrouded in ambiguity," we still refer to something definite behind the veil of uncertainty. Not so with privacy. Recall our discussion of invasive questions in chapter 1 and how the typical defense against them ("that's private") is not meant to preserve a secret or hidden piece of information. It just as often refers to something, like whether I am intimate with my wife or how much money I make, that can be inferred with a very high degree of certainty from public information. Rather, the response is meant to preserve a degree of social ambiguity or oblivion on that score—for others, but perhaps also for myself. This was Barthes's point when he said that "privacy is nothing but that zone of space, of time, where I am not an image, an object."[30] Psychologically, this ambiguity finds expression in something like the opposite of the oppressive, concrete self-identity involved in the experience of hiding.

An understanding of privacy as characterized by a tendency toward ontological and psychological ambiguity permits us to distinguish privacy's characteristic mode of self-relation from that of both hiding and solitude. Whereas hiding impales one on the fact of self-identity, and solitude enables one to split oneself into two mirror images, privacy, as we saw in the previous chapter, permits the self to come apart in a more fundamental sense opposed the imposition of self-identity. The forms of concealment we call privacy are, in other words, not for preventing the discovery of some particular piece of information or person, as hiding and secrecy are. Rather, privacy perpetuates a state of affairs in which there is no information, no fact of the matter, one way or the other—for the world at large, but also sometimes for oneself. This state of affairs is destroyed by surveillance or snooping *even if nothing hidden is discovered*, since a piece of information has been created by the discovery that I really did have nothing to hide. By contrast, before my privacy is violated (that is, while I still have it), the answer to the question of whether I have anything to hide would have to be something like "who knows" or "maybe, maybe not." Obviously this is not how it is with hiding and secrecy. The ambiguous response once again reflects the particular type of unknowing that privacy both protects and produces. This is also why it feels like an invasion of privacy when one is compelled to prove to the authorities or a lover that their suspicions of hiding something were unfounded. "Here, look through my phone," one might say to a suspicious lover, and though one consents to part with privacy in this case, one still feels it violated because the lover has treated one's privacy as if were instead secrecy or hiding, and because one was pressured to trade a condition of trustful unknowing for the certainty of information.[31]

We have said that a person in hiding is stuck to the fact of her self-identity, and that her plight consists in an inability to come apart from herself. This inability is not internal to her but imposed by the exigency of circumstances. Her experience is inimical to that of privacy in part because the experience of hiding is one in which she feels the pressure of her own facticity; the fixity of that part of her life which consists in information about who she is appears to her as a form of confinement, constraint, fate, or tether. We will better understand the opposition of this experience to that of privacy by borrowing from the literary scholar Kevin Quashie's distinction between two aural forms of concealment: quiet and silence. Quashie develops his theory of quiet to rescue and develop it as a cultural, aesthetic, and ethical category for appreciating Black experience and cultural production, and more generally as an avenue for human flourishing overlooked in an era focused on resistance, publicity, and expression. Quiet is a condition of interiority characterized by the "wild vagary of the inner life" and "the reservoir of human complexity." It is therefore "not consumed with intentionality" but stands in fundamental opposition to "what is calibrated and sensitive to the external world."[32] The forms of life and activity that Quashie identifies as calibrated to the external world include expressiveness and resistance, but also silence.

> Silence often denotes something that is suppressed or repressed, and is an interiority that is about withholding, absence, and stillness. Quiet, on the other hand, is presence (one can, for example, describe prose or a sound as quiet) and can encompass fantastic motion. It is true that silence can be expressive, but its expression is often based on refusal or protest [i.e., directed toward some other], not the abundance

and wildness of the interior. . . . Indeed, the expressiveness of silence is often aware of an audience, a watcher or listener whose presence is the reason for withholding—it is an expressiveness which is intent and even defiant. This is a key difference between the two terms because in its inwardness, the aesthetic of quiet is watcherlessness.[33]

The distinction between silence and quiet maps nicely onto that between hiding and privacy, as is perhaps belied by Quashie's description of the aesthetic of quiet not as the absence of listeners but *watchers*. Think of how a group of workers will fall silent when their boss comes into the room. Now imagine them sitting quietly absorbed in their tasks and thoughts. In the former, the boss is the cause of their reticence, as it were, which although autonomous is predicated upon the boss's not hearing whatever they had been talking about a moment before (which, if you've ever worked in such an environment, you know needn't be something salacious or potentially damaging). The boss hasn't silenced the workers, but their silence is directed at her, its necessary condition and ground. By contrast with the deliberate withholding of information in silence, Quashie's conception of quiet draws our attention to a state of enormous human potential that is grounded in its essential quality of unarticulated indeterminacy. With quiet there is much there that is not being said (everything, in fact) but at the same time there is no *thing* or bit of information or communication that is being withheld.

Connectivity and the Tethered Self

Let us return to the girl alone in her room. I chose this example not only because it is a central archetype of privacy but also because it is somewhat anachronistic. Today, this girl is very likely alone only in a physical sense. If she is anything like the typical teenager, she will probably have access to the Internet, probably through a smartphone that she uses to communicate with friends and strangers, bots and humans, to research and confirm, to catch up on current events, and to publicize her thoughts, pictures, and location. She is physically isolated, but obviously not in solitude. Is she still in private? Does she still have privacy?

We are inclined to think that she does because we are accustomed to base our judgment on certain visible, architectonic, or informational features of her situation—the conditions described in the "A has f but B" accounts of privacy above and reflective of the traditional association of privacy with the domicile and its property-based protections against trespass, nuisance, and physical invasion. However, one lesson we drew from thinking about how to describe the Franks' annex was that it is not sufficient for the girl to have privacy that she be alone, in her bedroom, unobserved with doors shut and curtains drawn. The discussion of hiding gives further reason for thinking that maybe the girl with the smartphone isn't properly described in terms of privacy, or at least that the language of privacy is less fitting than it was when she didn't have such a device.

If she were posting her thoughts and pictures on social media—that is, publicizing her life—it would be relatively easy to comprehend how her privacy could be diminished without it being actually invaded by a peeper or spy. The idea that privacy includes not just injunctions against unwanted perception but also norms against publicizing the parts of one's life that are properly kept from the public domain is an old one that has gone through several rounds of critique and defense.[34] But I will leave those arguments to the side for now except insofar as they touch upon the questions of connectivity and attention. For even if she does not publicize anything about her life, just in virtue of having a smartphone in her hands, the distinction between her condition in the room and the condition of the girl in the log begins to appear less sharp and more troubling.

One way to distinguish the pre-connectivity girl in her room from the post-connectivity one is by noting that the latter's attention is oriented toward publics in a way that resembles the tether of hiding more than before. In fact, "tethered" is just how Sherry Turkle describes the new relation of users to the technologies and networks that connect them at all times to one another and the Internet, especially the devices like smartphones that are "always-on/always-on-you."[35]

Turkle coined the phrase "the tethered self" to describe the new habits of mind, attention, and self-conception promoted by the technology and culture of connectivity. The phrase is meant to identify a technologically motivated change in subjectivity in just the same way that Paul Valéry spoke of the photographic revision of sight a century earlier. One of the new habits of mind Turkle identifies is the expectation that we can reach any person or piece of information at any time and any place, whose concomitant is the expectation that we ourselves can also be reached anytime, anyplace. The benefits of this arrangement are many and obvious, but it has a dark side, too.

Turkle's subjects, like those of similar ethnographies, report feeling "trapped and less independent" simply by virtue of being connected.[36] Although this is by now a fairly common sentiment, it has yet to lose its counterintuitive aura, since in many cases it is precisely this ability to be reached at any moment that is responsible for an increase in personal liberty. Think of the teenagers whose parents permit them to range farther from home on the condition of carrying a cellphone, or the adults answering emails and filling out spreadsheets on the beach. Most broadly, the sheer fact of constant connection to a panoply of media and discursive publics all constantly calling for our attention promotes the sense of never being alone, even when we are physically isolated. We are always "on call," "waiting to be interrupted."[37]

It doesn't matter where we are. We can still keep tabs on what others are doing, "stalk" friends and strangers, suffer FOMO, fret about how we are being perceived, and check in on the various conversations that are constantly going on in a multitude of discursive publics and inviting us to participate. "When media are always there, waiting to be wanted, people lose a sense of choosing to communicate," writes Turkle.[38] We feel the *pull* of the publics and technologies of connectivity, to which we often describe ourselves as being "tethered." The message one receives when feeling the pull of any tether, metaphorical or not, is that one's time and attention—that is to say, one's life—are not entirely one's own. In this, today's connected masses repeat the nineteenth-century view of photography's threat to privacy. Remember, for instance, the judge's description of the harm Paolo Pavesich suffered when his photograph was reprinted without his consent, noticing this time how the harm of enthrallment is tied up with questions of Pavesich's mental states and the abjection of his attention: "He can not be otherwise than conscious of the fact that he is, for the time being, under the control of another, that he is no longer free . . . conscious of his complete enthrallment."[39]

The dark side of the connected life bears a striking resemblance to the phenomenology of hiding. The hider's sense that she could be discovered at any moment is repeated in the experience of "waiting to be interrupted" and "always on call." Her compulsion to peek through a crack in the wall or a hole in the log mirrors the compulsion I feel to check my phone or email to see if anyone is trying to find me. The idea that what is hidden inclines to its own discovery is repeated in the assumption that under normal conditions in the era of connectivity, anyone can be reached—otherwise they are "off the grid," or lost, perhaps. To be sure, it is not obvious that the relationship between the tethered self is quite as abject as that of hider to seeker. Nevertheless, we can say that insofar as the tethered self is "at the mercy" of the media to which it is oriented, its condition more closely approximates that of hiding than privacy. Many of Donna Freitas's subjects express just this kind of abjection, speaking to her "of this strange, vicious cycle that being on social media regularly draws them into—the constant going online to check out what everyone else is doing and what your friends are up to, only to find out that this very form of 'socializing' ends up making them feel all the lonelier," which then compels them to start the cycle all over again.[40] In any event, this line of reasoning suggests an argument against a technology industry that seeks to make its products as addictive as possible—an argument in the name of privacy, in addition to the various other objections to the deliberate promulgation of addictive products.

Recall the phenomenology of hiding: the sense of being trapped, dependent, always anticipating an interruption from without and therefore with one's attention constantly being pulled out toward others who might be looking for you, perhaps by searching for signs testifying to your presence, the choices you have made, perhaps compiling a view of your psychology and character via those traces left behind in public. This sounds very much like what Turkle's and Freitas's subjects lament, if not in the admittedly counterintuitive language of hiding. But the compulsion one feels in the vicious cycle of attending to various virtual publics, to check every few minutes to see how many "likes" a post has gotten, fundamentally resembles the compulsion of the hider to peer through a chink in the wall or hole in the log to see if she is about to be discovered.[41] A more modest claim is that the two experiences are alike at least in the same ways that distinguish hiding from privacy. The tethered self is never really alone, hence the difficulty of achieving solitude with a smartphone in your pocket. But it isn't really present, either. "My mind is elsewhere," we complain.[42] This inability to be present can undermine all sorts of things: intimacy, friendship, comprehension, but also the ability to be with oneself or another in the deeper, unstructured sense discussed in the previous chapter.

The tethered self is, as one of Freitas's subjects put it, always conscious of "a vast potential audience [that] is out there waiting—waiting to be entertained but also to pounce when they see something they don't like."[43] It is revealing how the "vast potential audience" attending to the tethered self differs from spectators at a performance or even the regular readers of a newspaper column. It would be pathological for a performer to feel that they were on stage at home as well as beneath the proscenium, or to imagine that their audience was always out there waiting for them. It would be gross narcissism for the newspaper columnist to imagine that his causal conversation was addressed to a general public in the same way as his written words. By contrast, the tethered self's sense of the online audience is not confined to acts of publication but is ever present.[44] It does not matter whether the tethered self understands itself as a performer. It never, or rarely, disconnects from its potential audiences; they follow it out of the theater and into the street, the home, and, famously, the bathroom stall. This sense of an audience always waiting, that never leaves or goes home, is the obverse of the uncanny sense of appearing in public even when one's body is in private, which the photograph engendered in the nineteenth century and which the technology of connectivity has replicated and intensified in our time.[45]

The analogy with hiding finds further support in the development of the Internet into an easily searchable, public repository for the myriad digital traces we leave behind simply by living in the digital age. The sense that others, both known and unknown, are inferring from our traces what we are like very much resembles the experience of being tracked or run to earth, of one's self waiting to be discovered whether one likes it or not. Hence the anxiety that one has not been careful enough covering one's tracks, both characteristic of the hider and her orientation to the world beyond her concealment and a frequent refrain in studies of the connected life. The ubiquity of digital "tracking" in all domains of contemporary life acquires a darker aspect in this context. "It's like somebody is about to find a horrible secret that I didn't know I left someplace," is how one seventeen-year-old put it.[46] This leaves social media users feeling like they are "constantly engaged in a dance of minimizing 'risk' and avoiding 'danger.'"[47] Of course, this isn't just any sort of risk and danger, but one specifically having to do with how the traces we leave in public may be found and interpreted by others as leading back to us—not only to our bodies, but also to our personalities, histories, and moral characters. As a mode of relating to oneself, this constant engagement in a dance of minimizing risk and danger is not so different from the fugitive who must always be mindful to cover his tracks.

Blurred Lines

But we should ask ourselves whether the possibility of interruption that comes from a smartphone is perhaps not all that different from a landline phone or a knock at the door, which would mean that the girl in her room with her mind on Twitter and Reddit is basically engaged in the same sort of activity as her pre-digital counterpart when she conversed with or thought about others beyond the bounds of her privacy.[48] I think most of us have probably lain in bed at night obsessing over some misstep or silly comment, and I am sure that this nocturnal ordeal is much older than the Internet. Is this not a similar experience to those which we have described as inimical to the experience of privacy?

One glaring difference between the landline telephone (or the girl's mother calling her name through the bedroom door) and the smartphone is that the former are such close analogs to face-to-face conversation that they are practically extensions of it. Although a ringing phone can interrupt one's silence, sleep, or solitude, the invitation it offers is to connect one-to-one. It is an invitation to conversation, which is why it is a violation, so sharply felt among teenagers, when parents listen in on the line, and why it is a betrayal much sharper than gossip or broken confidence for one friend to trick another by having a silent caller listen in via "three-way calling." By contrast, the Internet puts us in the position of instantaneous publication, talking with strangers, reading or watching the myriad lives of others across the globe, and much more. These are activities that have traditionally belonged to the realm of publicity.

Whereas picking up the phone seems most closely related to letting a friend into her room to have a conversation (she "accepts the call"), posting texts and photographs, visiting various sites, stalking, and so on are activities that up until the digital age had taken place in public. Even the language we use to describe activity online is that of going out, not inviting in: we go to websites and visit them; we follow strangers; we post messages where they will be seen by friends or the general public. This distinction is strengthened if we compare the sort of connectivity that comes from a smartphone with that of television, radio, and newspapers, for those previous modes of publicity that flowed into conventionally private space were not open to the participation of all. Whereas a good radio program, like a good book, might transport one to distant lands, the virtual publics of the digital realm are spaces where individuals can instantaneously participate and therefore, in a very real sense, be.

The reality of privacy, as distinct from the reality of physical walls or the ocular fact of being unseen, is the reality of a social phenomenon, which means that it consists largely in how we understand and value it and what practices, norms, laws, and so forth we develop on the basis of that understanding. As a practice or set of values, privacy can be diminished or changed simply by our forgetting, or reimagining, the particularities of its boundaries. One far-reaching consequence of our constant connectivity is that the traditional lines separating the public from the private no longer appear so clear. If I am always reachable, I begin to lose the sense that there is a place where I can take myself apart from all others, that there exists a time and space where I can be alone, off the clock, off the grid. When such diminished expectation is shared among members of a society it becomes a social fact, and the lines that had once separated the different spheres of life begin to blur and become confused. This what Bernard Harcourt blames for "the mortification of the self" and the decline of what he calls "the liberal ideal."

> The liberal ideal—that there could be a protected realm of individual autonomy—no longer has traction in a world in which commerce cannot be distinguished from governing or policing or surveilling or just simply living privately. The elision, moreover, fundamentally reshapes our subjectivity and social order: the massive collection, recording, data mining, and analysis of practically every aspect of our ordinary lives begins to undermine our sense of control over our destiny and self-confidence, our sense of self.[49]

That something like this is underway appears in the increasing resemblance of the experience of private life to hiding. The view may be bolstered by comparison with the philosopher Jeffrey Reiman's sociological account of privacy. Privacy for Reiman is a complex "social ritual" of significant psychological and social-ontological value. "Privacy is an essential part of the complex social practice by means of which the social group recognizes—and communicates to the individual—that his existence is his own."[50] On Reiman's view, privacy confers "moral title to my existence" and makes "it possible to think of this existence as mine" in several ways, though most relevant here is the following.

I know this body is mine because unlike any other body present, I have in the past taken it outside of the range of anyone's experience but my own, I can do so now, and I expect to be able to do so in the future. What's more, I believe—and my friends have acted and continue to act as if they believe—that it would be wrong for anyone to interfere with my capacity to do this. In other words, they have and continue to treat me according to the social ritual of privacy. And since my view of myself is, in important ways, a reflection of how others treat me, I come to view myself as the kind of entity that is entitled to the social ritual of privacy. That is, I come to believe that this body is mine in the moral sense.[51]

As with the body, so too the moral ownership of one's life in the broader, ethical sense of self-conception and action.

Hiding is one type of experience that undermines my sense that my life is my own, for when I am in hiding I cannot take my body, myself, where I would like to, to show or share it with others. But the matter is much more diffuse in the digital age, when the technology and social expectations of connectivity mean that we are "never alone" and constantly sought by the world of work, social relations, and online publics. Reiman's description of privacy as a social ritual gives us reason to believe that generations who do not have the experience of taking oneself beyond the reach of others, or have it in sufficient quantity, may begin to feel as though they do not have moral ownership over their lives. Their sense of moral entitlement to lead their own lives might be diminished compared to generations reared amid more fulsome practices of privacy. To whatever degree, this is a serious injury. Such people would, no doubt, begin to feel alienated and helpless, much like how one forced into hiding feels. Perversely, they might also find the experience of being alone and disconnected from their technology to be alienating and frightening. This is, in fact, a common experience today.

One assignment I give in my college seminars on privacy asks students to take a solitary walk without any of their usual connective technology. Students report feeling anxious, uncomfortable, adrift in the world and, with unexpected frequency, "sus" meaning "suspect" or "suspicious," which is to say they view themselves through the internalized gaze of social norms as someone who has something to hide. The phenomenon is also reflected in social-scientific studies of the "tethered self," whose subjects describe not only feeling addicted to the new technology but have come to consider it an essential part of their bodies and subjectivity. For instance, one of Freitas's college seniors describes the common feeling that when he doesn't have his phone on him, "that's like me missing my heart or missing my brain."[52] These studies are replete with examples of the principle that the tools and technologies which become essential for everyday life, especially those affecting communication and the knowledge of others, have a significant effect on how we understand ourselves and therefore promote the development of new characteristics of human subjectivity.

The constant orientation to publicity enabled and encouraged by the technologies of mobile connectivity has negative effects for individual well-being and, as I will claim in just a moment, for public and democratic life too. I recognize that there is cultural variation on what constitutes over-orientation versus a healthy relation to publicity. My intent is not to prescribe rules or even standards but to bring to attention that there is a normative range here, with dangers at each extreme. As with too much attention to publicity, too little attention to the common world is itself an obstacle to individual well-being and democratic flourishing. Different communities and individuals can draw the bounds of what is acceptable in different ways, but if we fail to recognize that there is normative pressure here, from both sides as it were, then we are ill-equipped to make an informed decision.

Connectivity and Alienation

The idea that the era of mobile connectivity is one in which previous distinctions between the public and private are no longer so legible or reliable links the puzzle of hiding in private to a dark irony of the digital age, which is that an unprecedented increase in interpersonal availability coincided with a spike in loneliness and social isolation. We are more connected than ever, yet we feel increasingly disconnected. This came as something of a surprise and seemed to contradict years of public enthusiasm for the digital era's novel opportunities for communication and finding like-minded others. It seemed plausible that if, as Robert Putnam famously argued, increased alienation and personal despair came from a decrease in the "bridges" and "bonding" once provided by civic organizations, then the Internet might give postindustrial societies a second chance to find meaning by finding one another.

Yet the era of online connectivity coincided with new heights of loneliness and social alienation. The explanation for this is, no doubt, complex and overdetermined. Part of it has to do with the displacement of valuable forms of public, embodied sociality by virtual spaces of informational exchange and commerce. I want to suggest that a diminishment of valuable forms of privacy—that is, their shift to something more akin to hiding—and the loss of legible boundaries between public and private have also contributed to the phenomenon.

The importance of legible distinctions between realms of life was a recurring theme for Hannah Arendt over the course of her life. Consider a passage from her essay "The Crisis in Education."

> These four walls, within which people's private family life is lived, constitute a shield against the world and specifically against the public aspect of the world. They enclose a secure place, without which no living thing can thrive. This holds good not only for the life of childhood but for human life in general. Wherever the latter is consistently exposed to the world without the protection of privacy and security its vital quality is destroyed.[53]

One manner of being exposed to the world consists in being seen and heard by others. Another is the one we are concerned with here: of having too much of it streaming in. This is part of what I take Arendt to mean by the world's "public aspect," which is to say its face or appearance—in other words, its publicity. For Arendt, the public was the realm where individuals and collective works gained their objective reality via the simultaneous, reciprocal action of seeing and being seen. The Arendtian public therefore requires a pair of necessary conditions: a certain type of physical space and the perceptual and discursive activity that turns a plaza into a public square. We will have much more to say about this in chapter 5, so for now notice that the first condition relies on one meaning of the word public—that of being open to all and unhidden—whereas the second refers to what might best be called "publicity."

Publicity refers to a set of activities, expectations, norms, and channels of communication and knowledge, all centered upon the condition of seeing and being seen, hearing and being heard, but it's a bit more capacious and ambiguous than that. Like a climate, it is easy to identify but hard to delimit with precision. But it is this condition of publicity to which Arendt is referring when she writes of the aspect of the public world that can "invade" or "expose" the private life without trespass or destroying the four walls of the physical home. When, in *The Human Condition*, Arendt laments the rise of "the social," a public-private chimera that flows into and corrupts both the public and the private, it is just this sort of publicity that invades and corrupts the private while leaving the epistemic barriers of a house's walls untouched.[54] Since Arendt's focus is on the value of public life, she is not explicit about the mechanisms of how the entrance of publicity into private life threatens to destroy its vital quality, although her discussion of loneliness and solitude, along with the preceding discussion of hiding and privacy, begin to show a way.

For Arendt, isolation was necessary for an individual to act and to find a meaningful life among others, but it could also be destructive of those ends.[55] The sustaining form of isolation is solitude, which, as we have already seen, she understood as being alone in order to be "together with oneself" in preparation for returning to the public realm of action and meaning. This condition is famously destroyed when, in Wordsworth's epoch-making phrase, "The world is too much with us."[56] With loneliness, it is just the opposite: one has "the experience of being abandoned by everything and everybody,"[57] an experience that is often felt, and most painfully, in public, when one is surrounded by others. Solitude becomes loneliness when people are no longer able to be by themselves, in other words. The inability to be alone with oneself, along with the constant connection to others, is one of the hallmarks of the tethered self. This is the experience of those students of mine who find it uncomfortable or frightening to be left undistracted with their own thoughts. Arendt, like Reiman, gives one view of why this might lead to a feeling of impotence or alienation: "In this situation, man loses trust in himself as the partner of his thoughts and that elementary confidence in the world which is necessary to make experiences at all."[58] By contrast with her view of the self-sustaining experience of solitude, loneliness leads to feelings of irrelevance, impotence, and, for Arendt, paves the way for the willing embrace of totalitarian politics.

The view of privacy we have been developing gives another explanation, which we will expand in the rest of this book by connecting privacy and oblivion to the production of trust, depth, and meaning in human life. Another source of loneliness and isolation, as we have seen, is the tethered experience of hiding. One is fit "into the iron band of terror even when he is alone. . . . By destroying all space between men and pressing men against each other, even the productive potentialities of isolation are annihilated."[59] Who is more alone than a person hiding, notwithstanding her constant connection to the world beyond her hiding place?

4

MEMORY AND OBLIVION

Up to this point, our exploration of privacy has taken place in the present tense, so to speak. That one has privacy or is in private means that right now, the living, breathing individual is enjoying whatever benefits one gets from privacy, be it under the basic description of epistemic barriers, the repose of unaccountability, or the psychological freedom of being untethered. Of course, these perspectives on privacy may be said to be future-directed, as well, since we don't just seek privacy to prevent acts of unwanted perception but also to avoid the unwelcome consequences that might flow from them. Yet knowledge of ourselves and others is not confined to the present or the future. In fact, most of what we know about anybody, including ourselves, concerns the past: living memory and historical records. The importance of present- and future-directed barriers to knowledge, and the interests we have in those forms of oblivion, invite the question of whether we may have similar interests with regard to the past. Indeed, the language of oblivion seems most naturally germane to the sorts of epistemic barriers we find in the realms of memory and history. The word *oblivion* itself derives from the Latin for the act of forgetting, the state of being forgotten, and the willed pseudo-forgetting proscribed by laws of amnesty, and it lives on in Romance-language words for forgetting (from *oblivisci* we get *olvidar*, *oublier*, and their derivatives). The Spanish and French terms for what in English we call "the right to be forgotten"— *el derecho del olvido* and *le droit à l'oubli*—may be translated as *the right to forgetting* or better: *the right to oblivion*.

Recent widespread global activism in favor of a right to be forgotten makes perhaps the most explicit case for the value of oblivion in the information age.

Coming to grips with the moral vision motivating that activism, along with other arguments for normative limits to memory, commemoration, and the documentation of human life, will help to fill out the account of oblivion that we have so far been developing through an examination of privacy. We will approach the subject from three points of view. The first concerns the importance of forgetting for individual well-being. The second uses debates over the right to be forgotten to understand how the practices of so-called collective memory connected to the Internet's global archive may be said to harm individuals. And the third examines the perennial privacy concern over the increasing documentation of human life as such, what today we would call "datafication" but which has been a consistent concern of the public moral discourse on privacy since its origins in the nineteenth century. Each of these perspectives confirms what we have been arguing so far: that the possibility of being obscure to others and oneself plays a vital role in human life, past, present, and future.

The Unendurable Precision of Being

Recent decades have seen a surge of interest in the value of forgetting. It is hard to imagine that this is not a reaction to the culture and technologies of the information age, above all the development of the Internet. Books on memory published in the last several decades, even scientific or medial works concerned with its neurobiological aspects, rarely avoid mention of the World Wide Web. The advent of the Internet as a seemingly limitless, ubiquitous, and searchable archive seemed to many to mark a rupture in knowledge practices similar to that of the invention of photography and the rise of mass media recounted in chapter 1. Oliver Wendell Holmes's limitless enthusiasm for the photograph seems to verge on prophecy when he writes, in 1859, that "The time will come when a man who wishes to see any object, natural or artificial, will go to the Imperial, National, or City Stereographic Library, and call for its skin or form, as he would for a book at any common library."[1]

Yet for all the continuity of the long information age, it would be foolish to ignore the real novelties that the Internet introduced into everyday practices of knowledge. For one thing, the detailed recording of ordinary life that had once been the fate of political dissidents, criminal suspects, and celebrities has not only become the common lot but is today conducted with a degree of granular precision that the chroniclers and secret police of prior eras could have hardly imagined. Whereas for millennia the idea that "even the very hairs of your head are all numbered" required the existence of a deity to make any sense at all, now your Fitbit knows how many times your heart beats in a year, your phone knows how many steps you take and where, and predictive algorithms seem to know what you will want, think, or type before you do.[2] The possibilities for publication and publicity are also radically enlarged. Now anyone with an Internet connection has the power to enter information into a global archive, which we can all search, fast as thought, from anywhere in the world. This development gave rise to new practices and expectations around getting to know one another. Seemingly overnight, and almost without notice, we all became amateur private detectives, able to dig up information about people we have never met and to learn about the lives of private citizens that otherwise would have never touched ours. It is still incredible to me that in only the past five or ten years, the common usage of "stalking" to describe an obsessive attention to another's information or person has gone from denoting a serious and highly stigmatized moral infraction to the now utterly common, morally neutral activity of scouring the Internet to see what can be learned about someone.

To be sure, there are also many nontechnological developments that have contributed to the growing interest in forgetting. The development of the Internet coincided with, among other things, advances in the science of the human brain and a cultural shift away from Freudian psychoanalysis to other psychological paradigms that view forgetting as a possible route to well-being rather than the problematic work of repression. The politics of "never forget" that appeared as a self-evident moral imperative after the horrors of the Holocaust became complicated by less worthy practices of commemoration, like ethno-nationalist "lost causes" and bloody grudge-holding. We will have more to say about all of this shortly, but for now I wish to draw our attention to two things in particular. First is the astonishingly wide variety of perspectives and disciplines—from books of poetry to papers in peer-reviewed science journals and seemingly everything in between—that argue for the value of forgetting. Second, and even more astonishing, is that for all their wide variance in method and claims, these works more often than not converge upon a most unexpected point of reference: a short story published on July 7, 1942, in the Argentinian newspaper *La Nación*.

It is rare to encounter a work on forgetting published in the last twenty or thirty years written by a scientist, philosopher, physician, psychiatrist, linguist, lawyer, theologian, or anyone else that does not mention Jorge Luis Borges's short story "Funes, El Memorioso."[3] The story, whose title is translated into English as either "Funes, the Memorious" or "Funes, His Memory," concerns a boy named Ireneo Funes whose life is wrecked when he loses the ability to forget. It is a marvelous story by a master of the genre that exhibits the quality of insight into the human condition that we tend to associate with literary genius. Yet what is most interesting for our purposes is how in recent decades this story has floated free of its original context as a fantastical work of fiction and attained the status of something like a factual account of what a life would be like without the ability to forget. In both scholarly and popular works, Funes is presented as a case study or example whose predicament has more to do with the biological and agential reality of human memory than it does with Borges's vast oeuvre of fables and poems obsessed with remembrance.[4] Indeed, the story is often the only evidence offered in support of the claim that a life without forgetting would be intolerable, as if it were indeed authoritative on the matter.[5] One gets the impression that these writers must have encountered Funes in other works on memory and forgetting, since mention is never made of Borges's many other texts that reveal a life's worth of exacting and provocative thought on the subject.

It is surprising that Borges's story became a touchstone for understanding the value of forgetting to human life, not least because Funes presents an impossible case. For reasons I will give in a moment, our life can never be his life, notwithstanding the frequent assertions to the contrary and certain rare cases of human beings who approach Funes's asymptotically prodigious capacities of memory and perception. The usefulness of Borges's story as evidence for the actual workings of memory and forgetting is severely limited. The story is, however, highly revealing of the role that forgetting plays in our lives and culture, not because Borges "astutely foreshadows neuroscientific research,"[6] as more than one scientist has claimed, but on account of what the story's reception reveals about the society that treats it as if it were good evidence. The widespread acceptance and repetition of the story as expressing a truth about our capacities for memory and forgetting has made it a myth or fable of our time. It has become, in other words, the kind of fiction that tells a truth, not about what really happened or is actually possible, but about what people at a certain time and place find valuable. This makes the document an incredibly useful resource for anyone who wants to understand the ethical and political value of forgetting in the age of information. If we read Funes's story in the same way we read ancient Greek tragedies, for example, then understanding what makes Funes's story tragic will shed light on the value of forgetting in our time irrespective of whether it accurately describes how our brains work.

The story itself is quite brief. It recounts the narrator's three meetings with a Uruguayan boy named Ireneo Funes. Sometime between the first and second meetings, Funes is thrown from a horse and suffers the tragic injury that leaves him "hopelessly crippled." How crippled? "Now his perception and his memory were perfect."

> With one quick look, you and I perceive three wineglasses on a table; Funes perceived every grape that had been pressed into the wine and all the stalks and tendrils of its vineyard. He knew the forms of the clouds in the southern sky on the morning of April 30, 1882, and he could compare them in his memory with the veins in the marbled binding of a book he had seen only once. . . . He was able to reconstruct every dream, every daydream, he ever had. . . . Funes remembered not only every leaf of every tree in every patch of forest, but every time he had perceived or imagined that leaf.[7]

Funes learns Latin with only a dictionary and a volume of Pliny, which he can recite by heart after a single reading. Next come French, English, and Portuguese. He invents an idiosyncratic numbering system in which every integer up to at least 24,000 corresponds to a discrete image or object in the world. "My memory, sir," he tells the narrator, "is like a garbage heap."[8] His life is clearly meant to be tragic in something like the classical way or in the didactic mode of those folktales in which someone acquires a tremendous power that turns out to be a life-wrecking curse, which explains why readers naturally feel compelled to draw conclusions about the moral of Funes's story.

On the surface, Funes's tragedy reflects a judgment that our desire for a better memory, or to be rid of the blind spots that limit perception, is not without limits. This is not an idea unique to Borges: it is a prominent theme in several of Nietzsche's most read works and early classics of psychology like William James's *Principles of Psychology* and Théodule-Armand Ribot's *Les maladies de la mémoire*.[9] The connection of hubris and knowledge is the engine of history's most famous tragedy, Sophocles's *Oedipus Tyrannus*. The more pressing question is how to understand the limits of memory. Are they biological, social, or both? Are there normative limits in addition to empirical ones? Scholars of forgetting tend to draw the same conclusions: the ability to forget is necessary for the exercise of certain human capacities vital to agency and flourishing. Among these capacities are abstract thought, the use of concepts, the self-fashioning of personality, and the ability to act in the world. For the neuroscientist Scott Small, "Many of the story's passages about Funes describe a dominant cognitive impairment caused by his photographic memory: the inability to generalize—to see the forest for the trees."[10] In a book about the digital memory of the Internet, Viktor Mayer-Schönberger echoes the broader sense in which one's life would be wrecked if forgetting were impossible: "What Borges only hypothesized, we now know," which is that "through perfect memory we may lose a fundamental human capacity— to live and act firmly in the present."[11] The blatant category error of taking Borges's fiction as a "hypothesis" reveals once again the quasi-factual role the story has come to play in our time.

These interpretations misapprehend the nature of Funes's malady. For one thing, Funes *can* generalize. He uses abstractions and concepts all the time; it is just "*difficult* for him to see that the symbol 'dog' took in all the dissimilar individuals of all shapes and sizes," and "it *irritated* him that the 'dog' of three-fourteen in the afternoon, seen in profile, should be indicated by the same noun as the dog of three-fifteen, seen frontally."[12] His irritation requires access to the general concept *dog* and the abstract criterion of the fittingness of concepts, among whatever else. His successful use of the languages he learns and the deployment of the simile "my mind is *like* a garbage heap" likewise involve the exercise of abstract thought (and surely if one remembered everything one ever saw or thought, one would also remember the linguistic entities we call concepts). Nor has he lost his ability "to live and act firmly in the present"—he does it all the time, conversing and corresponding with the narrator, or when trapped with his thoughts during nightly bouts of insomnia. Like Funes, we too are reminded of past sights or experiences when, say, walking down the road or conversing with a friend, but those intrusions of memory do not mean that we are therefore not living and acting in the present. The only example Borges gives of Funes's torment has to do with his insomnia, which is described as the boy's inability to detach himself from the world: "*Dormir es distraerse del mundo.*"[13] At all events, being a good interlocutor or attending to our surroundings might be more difficult if our memories were like Funes's, but the tragedy of his life cannot be that certain aspects of it are more difficult, even as others have become easier, richer.

What the scholars of forgetting invariably overlook is that a perfect memory is necessary but insufficient for Funes's tragedy. Indeed, the story contains reference to other mnemonic prodigies whose lives, we are led to assume, were not thereby ruined. It is clear that the disaster of Funes's life stems from the combination of perfect memory with perfect perception. The difference with those other prodigies, and those clinical cases sometimes referred to as "real-life" Funeses, is that unlike Funes, their perception was still normal perception, limited by the natural capacities of the eye and attention. But Funes has no blind spots. He perceives everything whether he wants to or not. The merest glance and he has already seen every leaf of every tree in a forest. It is clear that it is the intolerable precision of Funes's knowledge of the world, and not simply his recollection, that makes his life "almost unbearable."

In the only scene where Funes's suffering is explicitly depicted, we see him tormented by a world in which everything has been accounted for and virtually nothing remains indeterminate.

> Babylon, London, and New York dazzle mankind's imagination with their fierce splendor; no one in those populous towers or urgent avenues has ever felt the heat and pressure of a reality as relentless as that which battered Ireneo, day and night, in his poor South American hinterland. It was hard for him to sleep. To sleep is to take one's mind off the world [distraerse del mundo]; Funes, lying on his back on his cot, in the dimness of his room, could picture every crack in the wall, every molding of the precise houses that surrounded him. (I repeat that the most trivial of his memories was more detailed, more vivid than our own perception of a physical pleasure or a physical torment.)[14]

Funes's life and the world around him turned oppressive and exhausting because he gained the power to know it all, all at once. Nothing is lost on Funes; he is overwhelmed by information, forced by fate to account for every vein in every leaf, every star in the sky. The relentless "pressure of reality" afflicting Funes recalls the stultifying "pressure of reality" that Wallace Stevens attributed to the torrents of information streaming into midcentury bedrooms over the radio.[15] From this perspective, the subject of the fable is more properly described as *knowledge*, for which both memory and perception are necessary. There is nothing in Funes's world that has not been turned into an object of knowledge, and there is no object of his knowledge that he does not know in exhaustive detail. Even the yet-unperceived can hold little hope, for Funes cannot doubt that he will know, in perfect detail, everything he encounters for the rest of his life. There may be new sights and experiences, new days, but each will inevitably be known exhaustively, without remainder. His is a life robbed of its fluctuant ambiguity, whether concerning his past, present, and future, or his self-knowledge and knowledge of the world beyond his head. Such ambiguity is essential for the humane value of living memory. The difference between Funes's perfect memory and ordinary living memory is that the latter is not fixed but constantly shifting and emerging from and sinking back into a mental background of oblivion. Experience with the mutability—the liveliness—of living memory is one source of the healthy sense that one's life is not fixed, that even in memory there is a quality of freedom, play, and potentiality. The exhaustive and unambiguous precision of Funes's knowledge of himself and the world, and the absence of any indefinite place or experience—the absence even of the hope for inarticulacy or oblivion—rob Funes's life of its quality of potentiality in a way similar to that discussed in chapter 1. This is why Borges would describe the precision of Funes's perception and memory as *intolerable*. And it is here, in its warning about the dangers of excessive and excessively accurate self-knowledge, that Borges's story most recalls Sophocles's play, reminding us once again of the Delphic wisdom: "know yourself," but also "nothing in excess."

We can imagine that a person in Funes's position may well come to feel that his life is closed down in a significant way. He is like a mirror—"a positively perfect mirror," as Poe described the photograph. It is not up to Funes what he will contain, just as it is not up to the mirror. Although all we know of Funes's experience is the little he shares with the narrator, only a fraction of which is conveyed to the reader, we can easily imagine it dawning on the unhappy prodigy that he shares with the mirror its lack of autonomy, its perfect inability to generate surprise, and such a total abjection to the world beyond his own body that it is as senseless to speak of Funes having a mental life of his own as it is to speak of a mirror having its own images. It would be understandable for anyone in such a situation to feel depressed or overwhelmed by a sense of life's futility. It certainly appears to have diminished Funes's understanding that his life is his and worth the trouble it takes to live it.

At the end of the description of Funes's insomnia, Borges gives a view of the boy's only respite from a world and life exhausted by knowledge. At night, when he lies in bed tortured by a kind of insomnia that is hard for us to imagine, Funes turns his mind's eye

> toward the east, [where] in an area that had not been yet cut up into city blocks, there were new houses, unfamiliar to Ireneo. He pictured them to himself as black, compact, made of homogeneous shadow; he would turn his head in that direction and sleep.[16]

Our interpretation of Funes's affliction is reinforced by the fact that he seeks solace in a form of oblivion, which he guards—almost, we want to say, holds sacred or keeps private from himself—in a continual effort to preserve the sole remaining element of ambiguity and potentiality in his life. Funes must believe that the houses which mark the limits of his perfect knowledge are not strictly imaginary—otherwise they wouldn't mark a real limit of an unknown region, but instead would be an act of naked self-deception. However, at the same time that these houses must have an objective existence in the world, they must also be something about which there is no information, no combination of details whose knowledge would exhaust their possibility and ambiguity. Black, compact, made of homogenous shadow. Against the exhaustion of perfect and unambiguous knowledge, of both the world and himself, Funes guards an unknowable reality and turns to it, as if in prayer.

Later we will see how privacy does not just protect oblivion but produces and confirms its reality by the visibility of its barriers to knowledge—walls, curtains, the social ritual of privacy, and the rest—just like Funes's zone of obscure houses. However, as far as Funes is concerned, no form of isolation or concealment offers any solace. He cannot sleep because he is "battered . . . day and night" by a "relentless" or "untiring" reality, which, in a phrase that recalls the phenomenology of hiding as much as Stevens's thoughts on the radio, prevents Funes from taking his mind from the world except with immense difficulty. Isolation offers no relief from this exhaustion precisely because his room is just as perfectly known as the world beyond it. Like one hiding, he is secluded but not enjoying the benefits associated with privacy. Like the insomniac who lies helplessly awake, relentlessly accounting for the missteps he may or may not have made the day before, Funes is unable to detach himself from his self and world. He cannot, as we put it in a previous chapter, let himself go or get out of his head.

Funes's life is exhausting because he knows the world, and himself, in exhaustive detail. This is the tragedy of Borges's story: the gain of an immense power of knowledge comes at the cost of something essential to human flourishing.[17] And this is why Funes's tragedy requires both perfect perception and perfect memory: if he had one without the other, the trap would hardly be so total. There would be either oblivion in retrospection (which Borges is careful to deny Funes by stipulating that after his injury even those pre-accident memories he had forgotten returned with a clarity and permanence equal to those formed in his new condition). Or there would be oblivion in perception, and Funes could live in the present moment and find some succor there. But since he can do neither, since the whole of existence is his to know and never to lose, it falls on him from minute to minute like a landslide, and he dies, as if suffocated by it, "in 1889 of pulmonary congestion."

A Right to Be Forgotten

An excess of memory is also thought to pose social and political hazards. However, it happens that most of the arguments about collective, social, or political memory are not really about memory, but rather the moral and political dimensions of historiography, commemoration, and related practices of collective identity formation and maintenance. The most influential of these arguments is probably Nietzsche's, which is also notable for advocating willful forgetting in terms of both individual memory and collective commemoration. Among other things, Nietzsche destroyed the illusion that history, to say nothing of its active commemoration, can ever be morally or politically neutral, arguing that because individuals are formed by the sort of society they are born into, "the ahistorical and the historical are equally necessary for the health of an individual, a people, and a culture."[18] Following Nietzsche, writers like David Rieff and Gabriel Josipovici have raised questions about the moral injunction to "never forget," which has migrated from its original context as a response to the Holocaust to be applied to all sorts of historical events.

On the one hand, the injunction to remember responds quite reasonably and justly to the fact that the organized horrors of the twentieth century were often accompanied by a deliberate intent to "eradicate the memory of what [the perpetrators] had done."[19] I know that I am not the only one to feel strongly the moral intuition, connected to reverence and awe for the deliberately disappeared, that we owe it to those whose lives were destroyed to preserve a part of them, even if all that remains are pieces of information in an archive. On the other hand, the idea that commemorating these events will prevent such horrors in the future—*never forget* as a means to *never again*—is not well supported by the evidence. In his book *In Praise of Forgetting*, David Rieff argues that collective remembrance has not only failed to prevent genocides in Cambodia, Bosnia, Kosovo, and Rwanda, or even to motivate meaningful intervention, but in some cases collective memory has stoked ethnic cleansing and racist nationalism. He recalls the "dismal fact there have been many occasions in the past when remembrance has been the incubator of a determination of a defeated people or group to secure vengeance, no matter how long it takes or what the human cost of doing so will be."[20]

Yet Nietzsche also suggests that the individual capacity of forgetfulness—without which "there could be no happiness, no cheerfulness, no hope, no pride, no *present*"—is not only a biological fact about our brains but also the result of certain social conditions.[21] He saw his own time as one that overvalued the historical, with the result that human beings had reason to envy the animals' ability to live only in the present while man "always clung to what was past; no matter how fast he runs, that chain runs with him."[22] Nietzsche's argument is Funes's implication: that human beings have an interest in being able to forget, to detach themselves from the facticity of the past. Might they also have a right to it?

This is the view of many who advocate for one of the novel rights of the digital age: the right to be forgotten, which gives individuals the power to erase or anonymize historically accurate information about themselves from the Internet. Not just any information, though, but only that which is "outdated," "inadequate, irrelevant or no longer relevant, or excessive."[23] Try asking a historian at what point information becomes "irrelevant" or "outdated," and you will see how much this right expresses a moral idea like Nietzsche's stance against excessive historicity. Advocates of the right to be forgotten are—no surprise—drawn to Borges's fable because it "seems to describe well the danger faced by a society that is no longer able to forget, such as today's dominated by the Internet and the web."[24] The phrase "the Internet never

forgets" has become a truism for our time, as has the idea that this state of
affairs presents a sufficiently serious danger to well-being to necessitate the
recognition of new moral and legal rights. The Internet's "unlimited . . . storage capacity has gradually become a threat to the privacy of citizens around
the world, since like . . . *Funes el Memorioso*, Google does not forget."[25] The
idea is that we are all approaching Funes's condition (or something like it)
thanks to the invention of the Internet's limitless archive. In recent decades,
legislatures and courts around the world have responded by giving legal force
to a right to delete, anonymize, or obscure information about oneself online,
while belief in a more basic, fundamental moral right to be forgotten is even
more widespread.[26]

As with the formative moment of modern privacy at the end of the nineteenth century, the rapid expansion of public debates over a right to be forgotten directs our attention to a realm of value still in the process of being
elaborated and understood. One curious aspect of that process is the language of forgetting itself, an unusual and somewhat counterintuitive locution that some advocates have sought, without much success, to replace with
a more literal description, based in what the right empowers its claimants to
do: "the right of erasure." Perhaps there is some political pragmatism behind
this, too, since most people hardly desire to be forgotten with the totality that
the English phrase implies. Nevertheless, these efforts have met with limited
success, and the majority of public moral discourse about the right still draws
upon the terminology of forgetting.

Even more curious is the common idea that the right to be forgotten is
somehow also a right of privacy, since as rights, the two seem to have at
least as many differences as commonalities.[27] Since we are speaking here
about the common perception, let us leave the arguments of this book aside
for a moment and speak somewhat schematically. Whatever one's understanding of the right to privacy, it is typically thought to be concerned with
something (information, bodies, objects) that is not already freely available
in the public sphere, while the right to be forgotten concerns *only* information that is already freely available in the public sphere. Although we might
think privacy rights should be concerned with information that entered the
public sphere as the result of an initial privacy violation, it would upend
the long-standing understanding and usage of privacy to insist they should
also extend to information that was unproblematically publicized and freely
accessible for many years. In stark contrast, recent publications are assumed

to be newsworthy under the right to be forgotten and therefore beyond the reach of the right's legitimate exercise, which is explicitly restricted to information on the Internet that has been publicly available for so long that it can be said to be "irrelevant" or "outdated." The idea that rights to privacy could be restricted to certain kinds of information (other than "private") is anathema even to those who think that privacy is, in fact, for protecting information.

Further distinctions may be drawn, but let these suffice for the conclusion that we ought not to treat the two rights as reducible to one another. Once again the danger here is not merely conceptual but political. Advocates for a right to be forgotten should be wary of its association with a long-standing body of thought about privacy that has considered true information that was not publicized as the result of a violation to be beyond the scope of the right's legitimate exercise. Rights to privacy also risk being undermined—for instance, by the idea that some forms of such rights are legitimately restricted to "outdated" and "irrelevant" information.[28]

It is a mistake to equate rights to privacy with rights to be forgotten, or to think that one derives from the other. However, it would also be a mistake to ignore the common sense and strong intuition that there is some relation here. Indeed, the language of oblivion suggests as much. The two rights are, in fact, related, because they protect the same thing, albeit in significantly different ways.

The Story of K

In 1981, a German man was hired to help sail a yacht from the Canary Islands to the Caribbean. Out of respect for his right to be forgotten and to avoid calling him "the murderer," let us call him K. Tensions between K and the others onboard rose over the course of the transatlantic voyage until, following an argument over unwashed breakfast plates, K pulled out a gun and shot the yacht's owner and his girlfriend. They died, and K was apprehended and returned to Germany, where in 1982 he was convicted on two counts of murder and sentenced to life in prison. The crime was something of a sensation. It was covered in newspaper articles, long-form magazine pieces, a book, a documentary film, and eventually became something of a touchstone in the popular and scholarly writing on "yachting and madness."[29] K served thirty years of his life sentence and was released early from prison in 2002. Shortly thereafter, he typed his name into an Internet search engine. The search returned, among whatever else, three articles published in the magazine *Der Spiegel* in 1982 and 1983, which had been uploaded to the Internet a few years prior in the course of the magazine's digitization of its archive.[30] These articles mention K by name. In 2012, he sued *Der Spiegel* in an attempt to force it to delete or anonymize the articles in its online archive, and in 2019, Germany's highest court took K's side in the matter on the grounds that the articles violated his right to be forgotten. The court further agreed with K that the information connecting his name to his crimes on the Internet—but only on the Internet—threatened his fundamental interest in "the basic conditions enabling the individual to develop and protect their individuality in self-determination."[31]

The standard understanding of how K may be harmed by the presence of those articles on the Internet draws on an idiom of tethers and shackles that recalls the nineteenth-century descriptions of privacy harms, the "tethered self" of mobile connectivity, and Nietzsche's analogy between the inability to forget and the inability to outrun one's chain.[32] The presence of historically accurate personal information online is said to pose a grave risk to individuals because it threatens to "forever tether us to all our past actions, making it impossible, in practice, to escape them."[33] One is said to be "shackled to the past," "frozen in time,"[34] and "a prisoner of [one's] recorded past."[35] The aspect of the Internet responsible for shackling people to their pasts typically laid at the feet of the cliché that "the Internet never forgets."[36] In the words of Jeffrey Rosen: "The fact that the internet never seems to forget is threatening, at an almost existential level, our ability to control our identities."[37]

But this cannot be right. Although it is true enough that the Internet never forgets, it doesn't remember, either. This is no less true for paper archives, which have so far proved to be a more durable form of information storage than the rapidly obsolescing hard drives on which the Internet's information is stored.[38] Like other forms of archive—indeed, like the much older information technology of writing—the Internet is a vast historical resource that can supplement or aid memory but is not identical to it.[39] In strictly mnemonic or historiographical terms, the Internet is no different from offline archives. The massive difference, of course, has to do with accessibility.

For the first time in history, a great and ever-increasing proportion of humankind has a publicly available documentary record providing all sorts of information about them—not just where they live, what they do, what they have achieved or committed, but also records of their thoughts, images from their daily lives, and the public interpretations and commentaries from many others, all of which is indexed and easily searchable. Yet it seems clear that without ubiquitous access to this archive, the Internet would hardly have posed the novel threat to well-being claimed by advocates of the right to be forgotten. If we had to go to our local library—like Holmes's City Stereographic—to access it, for instance, the effect on the structure of our ordinary practices of personal knowledge would have been comparatively minor. Without ubiquity of access, the Internet, no matter how powerfully searchable, would have presented a marginal improvement on preexisting technologies of information storage and retrieval, like microfiche and CD-ROM, which were themselves marginal improvements on paper archives (if improvements they were). As occurred with the integration of photography into everyday life at the turn of the twentieth century, the ubiquitous availability of the Internet's easily searchable archive and its thoroughgoing integration into our quotidian practices of knowledge caused a shift in epistemic practices of personal relation at the turn of the twenty-first. It is this shift in practices and the new expectations it introduced into everyday life to which the right to be forgotten responds, and not merely the technological conditions that were necessary for it.

Enforced forgetting is notoriously counterproductive. We still know the name of Herostratus, not because he did anything particularly notable, but because Greeks in the fourth-century BCE were prohibited to remember him after he burned a temple in the hopes of achieving eternal notoriety. K's right to be forgotten obviously cannot make anyone who already knows about his convictions or the *Der Spiegel* coverage forget about them. All it does is empower him to change his state of affairs from one in which it is reasonably likely that new acquaintances would discover the article to one in which it is reasonably unlikely that they will. In other words, the right makes it such that new acquaintances will likely be *oblivious* to K's past or certain elements of it that appear online.[40] The right cannot guarantee that K will form new relationships that are not preconditioned by knowledge of his past, but it does give him the *possibility*. This possibility is what the right to be forgotten is actually a right to, which is why K sued online publishers and search engines rather than the individual people who remembered the articles or confronted him with them. This focus on the restoration of a certain possibility also helps us to understand how K may reap the benefits of the right even if he never actually meets anyone who doesn't know about the *Der Spiegel* coverage. If K's right to be forgotten is a right to the possibility of meeting someone who does not know about his past, but *not* a right to such meetings in fact, then we can more clearly identify the corresponding injury at which it is aimed. It is not literally impossible for K to form new relations un-preconditioned by the information in the articles. Nevertheless, the reasonable expectation that most people he meets will look him up at some point early on in the relationship means that, as long as the articles are there and connected to his name, K is deprived of the reasonable expectation of a certain amount of social obliviousness about his past, which in turn contributes a sense of his life going forward as being closed down in a morally significant way.

It is not possible to determine what others will think or remember about us. The world of others, too, has a right to make up their own minds about what we are like. Nevertheless, healthy agency requires the confidence that the direction of one's life is up to one and that it is worth the trouble it takes to direct it. If K has the sense that everyone he meets will know about his convictions ahead of time, and he can reasonably infer the sort of conclusions they will draw about what he is like (dangerous, crazy, etc.), then he may come to doubt whether he will ever be able to be otherwise than what the articles imply. Or he may lose confidence that it is worth the trouble of trying because the outcome seems determined in advance.[41] Without such confidence, the knowledge that that one's life is one's own is ambivalent at best. In fact, it is precisely the understanding that one's life is one's own and no one else's—that is, it is the integrity of personal identity—that underwrites the sense of being shackled to oneself or one's past. The belief that one's life is one's own but with no point in trying to live it according to one's lights is one of the more painful forms of estrangement from oneself and the world imaginable. We should expect one in such a situation to end up like Funes, for whom even the isolation of privacy offered little of its characteristic repose.

To the extent that K loses either or both elements of such confidence, he will be undermined in the domain of agency. K suffers this harm independently of whether he ever encounters someone who recognizes him as the guy from the yacht murders. If he never leaves home but still doubts that it is possible or worth the trouble to direct his life as he sees fit, in terms of agency he is equally worse off as if he had gone out and been recognized. What K has lost is not a set of actual possibilities—at least we don't know that, and anyway the right to be forgotten doesn't require that he demonstrate such a loss. Rather, he has lost confidence in his possibility of being different from himself or, in the language of chapter 1, belief in the potentiality at the heart of human personality and action. Michel Foucault expresses a version of this idea:

> The main interest in life and work is to become someone else that you were not in the beginning. If you knew when you began a book what you would say at the end, do you think that you would have the courage to write it? What is true for writing and for a love relationship is also true for life. The game is worthwhile insofar as we don't know what will be the end.[42]

The dread that one's future is written in advance is the malady of Funes transposed from past to future.[43] To be sure, it is not inevitable that someone in K's position will lack the confidence that it is still possible and worthwhile to be different from himself—that his life, in other words, is not already finished but still contains its inexorable quotient of potentiality—but it is not unreasonable, either. And if he were to feel this way, he would have a direct remedy in the right to be forgotten.

Think of the celebrity who cannot go anywhere without being recognized. It is not uncommon to hear a movie star say that she chooses to vacation in a certain spot because nobody there knows who she is. It is a relief to go unrecognized. What the ignorance of the locals gives her is neither privacy nor forgetting, but oblivion. It is something close to anonymity, although that does not fully capture what it is she seeks in a place where she's not ogled unknowingly or recognized as "that person from that thing, you know . . ." Rather, what the backwater provides is a quality of openness that her life typically lacks. When she meets someone who does not recognize her, it is up to her (as much as it ever is) what sort of conclusions the other will draw about her. That moving in such a milieu would be described as a kind of relief or respite is another testament to the importance of spaces and opportunities for the coming apart of personal identity.

But we might still want to be a movie star, all things considered. So think instead about the practice of criminal branding. What is morally repugnant about criminal branding is not primarily the temporary infliction of physical pain. Rather, what makes branding barbaric is the attempt to destroy the branded person's confidence that it is possible for him to ever meet another person without that meeting being conditioned by knowledge about his past. It is hardly less barbaric if the brand is administered under anesthesia or can be covered by one's clothes. The painless brand is a moral horror even in a society so highly tolerant that it presents no obstacles to one's life projects—except to becoming *the sort of person who was never branded*, which is exactly the point. This is because the criminal brand attempts to make the branded individual feel that his past is insuperable—in other words, that he is shackled to it.[44] As K does not need to meet with someone who knows about his past to suffer the relevant harm, neither does the branded one have to wait until someone sees it for him to have suffered from the harm and barbarism of branding. If painless branding is morally unjustified for this reason, then so too would be a state of affairs in which technological advances achieve the same result by different means. The analogy is much less far-fetched than one might think, for the branded person's past is not insuperable or "shackling" on account of anything done to his memory or the memories of others, but rather because of a technological intervention (albeit somewhat primitive) that undermines his reasonable expectation of a certain amount of social obliviousness.

This brings us to a deeper understanding of the harm of being shackled to the past. Being shackled is bad not only because of how it limits what is possible to do or achieve in one's life. It also injures in the message it sends to the shackled person that it is futile even to try exceeding the limit of his chain. The shackled person is worse off to the extent that he comes to see the direction of his life as futile as well as constrained. Like the literally shackled person, K loses a bit of the confidence that his life is his to direct or that it's worth trying. Surely he doesn't lose *all* such confidence, but even a minor loss in this area is significant and can spread to other domains of self-understanding. It is important to pay attention to the apparently small threats resulting from unexpected technological shifts in social life, lest they come to be taken for granted and lose their appearance as threats, which in turn would mean that we have lost a fuller view of the interest at stake.

To say that K has an interest in the confidence that his life is up to him is not to say that he therefore has a right to a particular psychic state, such that those around him would have a duty to make it so. However, he can have a right to the conditions that make such confidence reasonably available to him. As with all forms of confidence, the confidence that one's life is one's own is not a natural fact of human psychology but a response to certain social, material, and ideological features of an individual's surround. These features can support such confidence, but they can also undermine it. "Stones can make people docile and knowable," in Foucault's famous phrase.[45] This was the lesson of Bentham's panopticon, whose reformational force depended upon the architectural erosion of an individual's confidence that it is up to him to decide how to be. And it was the lesson of Foucault's genealogy of the new "physics of power" that Bentham's plans represented, which highlighted the many ways disciplinary power took hold over modern life through the increasing individualization and documentation of the self.[46] In the political domain, this physics of power manifests in the variety of ways that modern states have sought to undermine citizens' confidence that it is for individuals to decide how to be, from ubiquitous surveillance to social credit scores and reeducation camps. Advocacy for the right to be forgotten reveals a common belief that the ubiquity of indexed, accurate information about individuals can achieve a similar effect, albeit by means more diffuse and less extreme.

Conceived in this way, the right to be forgotten offers an even broader sort of protection than commonly thought, whose remit extends beyond remedying the individual harms of those who bring claims. The sheer existence of the right to be forgotten gives everyone in a society the reasonable confidence that if one day they were to feel shackled to their pasts thanks to information online, they would have a remedy to undo their chains. This is a public good of the right to be forgotten, which supports the sense of an open future in those who have not yet felt the need to exercise it but may someday want to, and thus it holds open a far broader sense of possibility in human life. Another aspect of this public good draws on the arguments about the value of oblivion. Like Funes's obscure houses, the right protects a degree of potentiality and oblivion in human affairs. It achieves this by its use, of course, but also through its expressive dimension, which sends the message that there is a realm of human value pitched against information and knowledge.

Here we return to the connection between the right to be forgotten and privacy. The idea that information circulating about oneself preempts one's opportunity to make a first impression echoes early complaints about photography and privacy, where one's image was thought to speak for oneself in ways, at times, and in places disconnected from one's body and will. One aspect of the harm Paolo Pavesich was imagined to suffer from the unwanted circulation of his photograph was an *awareness* that he is "under the control of another, that he is no longer free, and that he is in reality a slave without hope of freedom, held to service by a merciless master . . . conscious of his complete enthrallment."[47] The injury of the photograph's fixity returns in the Internet age as the fixity of information about one's past. As Barthes lamented the replacement of the fluid, protean self with the immobile solidity of the photographic image, the fluidity of human memory and its collective manifestation in the form of meeting strangers is fixed in the deadened, unambiguous form of information. Earlier concerns with the fixity of the photograph likewise echo in the accounts of the Internet's "iron memory" sticking us to our pasts.[48] The confidence that we could be different from ourselves if we wanted, based in a social structure that makes such confidence reasonable and equitably available, is a fundamental condition of healthy agency and human flourishing. This is not an idea confined to contemporary advocacy for the right to be forgotten but has animated more than a century of public moral discourse that pits the value of privacy against the increasing documentation of human life in the information age.

An Ethics against Fixity

Moral concern about the accumulation of personal information in archives has been a constant feature of the public moral discourse around privacy since the inception of the modern value. As the historian Sarah Igo observes, "Privacy was the language of choice [in the nineteenth and twentieth centuries] for addressing the ways that US citizens were—progressively and, some would say, relentlessly—rendered knowable by virtue of living in a modern industrial society."[49] In addition to photography and newspaper mass media, the demise of privacy was declared in response to the issue of driver's licenses and passports; the development of fingerprint registries, personality tests, the US Social Security registry and numbers, credit agencies, and corporate computer data banks (described at the time as "record prisons"); the storage of mortgage and tax records, marriage and death certificates, elementary school records, hotel registers, and much else.[50] These were complaints against "big data," albeit not in those terms.

This parade of moral panics reflects a long-standing and widespread ethical concern with the increasing documentation of human life. Yet by comparison to prior advocacy against other information technologies, the proposals for a right to be forgotten appear like half-measures. For although those earlier complaints of privacy against information prefigured today's unease at the constant creation of information about the most miniscule aspects of our lives, they expressed a different understanding of both the problem and its solution. What bothered nineteenth- and twentieth-century opponents of "big data" was the entry of more and more of human life into documentary records *as such*. What they demanded in response was not more control over the information or the ability to freely direct it where they wish, but rather for that information not to be created in the first place—or, second best, for it to exist no longer. As Igo writes about the early twentieth century United States, "Life insurance and credit agencies were just two of the powerful entities driving the creation of what we would now call 'personally identifying information,' the sheer creation of which constituted a 'terrible invasion.'"[51] Nearly a century later, Bernard Harcourt laments the loss of a more demanding approach to the threats of big data in his critique of neoliberalism's drive to turn as much of life as possible into the type of information that can be used to calculate, predict, and pin down: "Quantifiable material interests took precedence over spirituality, lessening the hold of the more ethereal concepts, such as privacy, human self-development, autonomy and anonymity."[52]

Although today Harcourt's view—like mine, in fairness—is somewhat marginal, it hasn't always been this way. For more than a century, the common thinking about privacy maintained just this sort of opposition and mutual exclusion between the parts of human life that were private and those that were entered "on the record."[53] Examples throughout this book and our everyday lives reflect the fact that this idea still animates many of our intuitions about privacy, even if the explicit theory of its value has been largely exchanged for a focus on control, access, and integrity that takes for granted the compatibility of what is private and what can be on the record.

The language of invasion in anti-documentation advocacy recalls the analysis in chapter 1 and suggests an explanation for how anyone could think that the information held by an insurance agency could be invasive, especially since at the time such information would have been only what the policy holder had volunteered, complemented by actuarial tables and statistical data. It was invasive because it made some fact of the matter about one's life where before there had been none. To be sure, we might not consider this harmful in the same way as K's situation or the Voyeur's Motel. Yet even if we do not think that this kind of privacy invasion is the sort of thing we ought to have rights against, it nevertheless directs our attention to an area of moral concern about human well-being in the information age. And although the complaint against insurance agencies may seem fanciful today, it is not so different from contemporary privacy concerns about insurance companies monitoring how we drive, what we buy, how much time we spend on the Internet, and how often our phones run out of battery. Citizens still find such programs "extremely intrusive," even when they are voluntary.[54]

Paying attention to this long tradition of advocacy for privacy against information emphasizes once more the misguidedness of understanding privacy, the right to be forgotten, and other values ranged against the conversion of human life into information as based in the importance of controlling one's information, or maintaining the integrity of different personas in various social contexts, or the projection of a particular self-image. These views are not equipped to problematize the sheer increase in documentation about human life, or to comprehend the "intrusiveness" of consensual tracking in public, which means they fail to reflect more than a century's ordinary use of the term they seek to explain. As a practical matter, this deficiency also leaves them relatively toothless to confront the relentless drive of surveillance capitalists and states to turn as much of human life as possible into information and data.

There are several reasons to be concerned about the increasing documentation of human life. The data could be misused. Precise information about us, along with the algorithmic curation for which it is used, could be used to motivate us to abstain from voting or to view a certain group of people negatively, or to stalk or blackmail us.[55] As Latanya Sweeney and others have shown, anonymized information (like medical records) can be linked to discrete individuals with shocking ease and a handful of datapoints.[56] Governments can use information collected by corporations to keep tabs on citizens and quash dissidents; police may purchase personal information from data brokers, circumventing the warrant approval process they would have to go through if they were to do the surveillance themselves.[57] If information about our behavior is harvested and commodified and resold on the secondary market, then we may also have a complaint about the exploitation of our labor.[58] If the possession of huge amounts of data is tantamount to the possession of political power (by way of its usefulness in predicting and influencing human behavior), and if the commoditization of data means that it will overwhelmingly be controlled by the wealthiest, then one might reasonably worry about democracy sliding (further) into plutocracy.[59] And so on.

But what's bad about these possible consequences, bad as they are, does not have anything especially to do with privacy, but rather with the badness of being manipulated, working for free, and living in a plutocracy. They are also contingent. Although wisdom councils that the stockpiling of information about us makes its abuse not only possible but practically inevitable, it is not certain. (The logic of surveillance: collect everything you can, since you never know what may one day prove useful.) As far as bad uses of information go, the danger of it simply *existing* consists in its potential for abuse. This is what Carissa Véliz captures with her evocative metaphor of personal data as a "toxic" asset: it is "the asbestos of the tech society."[60] Asbestos is not dangerous until it is disturbed and inhaled, but since it is not strictly necessary for human life and so dangerous, we have decided to ban its use entirely. Véliz argues that we should do the same for the data economy.

We might also think that the documentation of much of what we do, say, and buy can exert a kind of objectionable pressure on our behavior. I don't need to be certain that my Internet searches are being recorded somewhere to hesitate before using keywords that might one day attract the authorities to my file. I internalize the surveilling documentary authority, and without meaning to or perhaps even realizing it, I monitor myself and adjust my behavior with respect to some standards that *might* or *could* be applied to judge my record of behavior. This is, in broad strokes, the famous view of panopticism and the disciplinary society that emerges from Michel Foucault's *Discipline and Punish*.[61] Foucault chronicles how the spread of documentary practices in the eighteenth and nineteenth centuries "lowered the threshold of describable individuality and made of this description a means of control and a method of domination."[62] The very idea that one's past can be *looked into* echoes the complaints about the doctor's scope that could see inside one's mind. But let's leave this aside with a "be that as it may," for although this effect is surely real, it has been so fully explicated that it does not need another airing here. It also fails to fully capture what we are after.

If I discover that my neighbor has been keeping a diary of my comings and goings, what I do in the garden, when we have friends over, what music or baby wailing she hears emanating from within our dwelling and when, and if I discover that she does this only for her amusement and will certainly never use her journal to harm me, I nevertheless feel that I've been affected in some way, although she has only recorded what anyone would expect a neighbor to perceive. And I reach for the language of privacy to explain this feeling. Indeed, the most natural way to describe her actions are that they are, somehow, *invasive*. I expect my neighbors to perceive me in these ways; such perception on its own strikes me as entirely innocent. Nevertheless, there is something untoward in my neighbor's behavior which, for reasons that should by this point no longer seem so counterintuitive, pertains to a sense of privacy. There is no concern that this information can be used to harm or manipulate me. Neither do her observations and record-keeping affect my freedom, or autonomy, or self-projection, or my ability to project different personas in different contexts. Part of my moral intuition in this case has to do with the neighbor's *scrutiny*, which we previously discussed as a form of attention concerned with accounting for the features and identity of a particular object. But the fact that she keeps a written record seems to add something to the equation, in the same way it would if I found out that she took pictures of me coming and going for her personal records.

Keep in mind that it was only in the course of the long twentieth century, in the United States and much of the rest of the world, that official records on all people, from birth to marriage to death, and in increasing detail (income, eye color, address, gender), came into being.[63] Foucault summarizes this shift in the practices of personal knowledge at the era of its inception as "nothing less than the entry of life into history, that is, the entry of phenomena particular to the life of the human species into the order of knowledge and power."[64] He puts into theoretical language what those who lived through these changes already knew. The expanding documentation of human life did not simply record preexisting facts for preservation. Rather, new methods of measuring people created new characteristics which they were then said to have. Each novel mode for creating information about individuals and each individual measurement constituted "the establishment of a truth"[65] where there had been none before, or at least where none had existed with the fixity, legibility, and communicability of documentary evidence.

This is not to say that there was no truth about ordinary people before the nineteenth century, that there was no information about their lives. Privacy concerns over the increasing documentation of human life were not focused solely on the creation of information about people where before there had been none; they also reflected a qualitative shift in the means by which those facts were constituted. The surge in *documentation* about persons reflects a technological transition from the malleable, fluid mnemonic practices of "living memory" to a documentary record characterized, like the photograph, by a degree of material fixity, portability, and an appearance of epistemic objectivity. Even at the turn of the twentieth century, citizens had a sense of human life as headed toward something like Funes's memory, if perhaps not yet approaching it. The world that, through the magic of photography, drew itself with "the pencil of nature" was also a world turned against itself, entering itself into the realm of fixed documentation by the legions of "Kodak fiends driving the world mad." Newspapers, too, seemed to represent a novel proliferation and dispersal of the "public record" which previously had been largely confined to police archives, a trend continued with and amplified by the development of the Internet, social media, and the search engine a century later.[66]

Our worries about the ethical consequences of the rampant documentation of human life find a sympathetic, provocative interlocutor in Foucault's last works.[67] Toward the end of his life, Foucault shifted his attention from diagnosing the various ways that the inescapable compound of power and knowledge manifested itself throughout history and instead began to ask whether there might be an ethics of self-knowledge that did not enmesh one in the dominating structures elaborated in his earlier works. If, in those earlier books, the production of knowledge about individuals was inseparable from the complex relations and strategies of power, the later works and lectures reveal an interest in the liberatory potential latent in certain practices of self-knowledge, especially those dedicated to its coming apart. Whereas previously he had argued that "between techniques of knowledge and strategies of power, there is no exteriority," now he asks "what would be the value of the passion for knowledge if it resulted only in a certain amount of knowledgeableness and not, in one way or another and to the extent possible, in the knower's straying afield of himself?"[68]

Foucault's move from one Nietzschean question—what is behind self-knowledge?—to another—what good is it to human life?—subordinates knowledge, like everything else in human affairs, to the greater question of well-being. This move led him to an ethics of self-knowledge concerned with self-detachment. The phrase he uses for this experience, aim, and value is "*se dépendre de soi-même*"—which can be translated variously as "to take one away from oneself" or to "untake oneself," "to release oneself from oneself," "to disassemble the self, oneself," and "to get free of oneself." It is significant that the phrase Borges uses to describe what was impossible for

Funes—"distraerse del mundo"—also means literally "to take oneself away," "to pull oneself apart" or "to create distance between the self and the world." A view of well-being concerned with the coming apart of personal identity is also what animates the understanding of the right to be forgotten as a right to undo the tether that binds one to oneself.

It is not hard to see why Foucault, like the others we have seen worrying about the fixity of identity, would be attracted to such an ethics of self-knowledge. Against the archive and the forces intent on assembling individuals by assembling information about them ("building profiles" in today's telling phrase, as if the information about us were equivalent to that emblem of individuality), an ethics concerned with individual freedom and agency would naturally find itself concerned with values, practices, and spaces dedicated to self-disassembly.[69] If the problem concerns the fixity of an individual in information, then the answer cannot be more or better information. Nor does it suffice to change or to contradict oneself, if those changes are entered into the field of documentation under the presupposition that they describe a singular, self-identical person who has changed over time. One is different, but one is still one.

Following in the vein of Foucault, if not his exact tracks, Édouard Glissant seeks to resolve the ethical and political questions of writing and identity with the value of what he calls "opacity," which has since been taken up by postcolonial, feminist, and queer theorists and activists as an ethics of the fluid self in an era captivated by the fixity of information. In opposition to the abjection of information, Glissant argues for the value of opacity as "that which cannot be reduced."[70] We may hear a silent "to information" ringing at the end of Glissant's line, for what cannot be fully known cannot be reduced. The artist Zach Blas, who designs obscenely globular masks that obscure the face from tracking software while making glaringly visible the obscenity of the technology's imposition on human life, reads Glissant as arguing for a personal and political "alterity that is unquantifiable, a diversity that exceeds categories of identifiable difference." Opacity, like Blas's masks, "therefore, exposes the limits of schemas of visibility, representation, and identity that prevent sufficient understanding of multiple perspectives of the world and its peoples."[71]

Blas is drawing from *The Poetics of Relation*, where Glissant opposes a politics and ethics of opacity to the "transparency" of a postcolonial critique fixated on the question of identity in difference. Glissant recognizes the achievements of this movement while arguing that the emphasis on difference merely inscribes another kind of fixity for postcolonial subjects. A self, a personality, or a politics so conceived is abject in the same way as the subject of those theories of privacy which understand the value of concealment as derivative of projecting a certain self-image in public. The conviction that human life is altered when converted into information weighs against many claims of "clean hands" made by the industrialists of today's data economy, who rely on the fact that the aspects of our lives they monitor and record are not secret or concealed, but often take place in plain view.

To be sure, there are differences between the accounts of obscurity espoused by Glissant, Foucault, and this book. Nevertheless, they converge around a point that has also been central to the public moral discourse about privacy from the beginning: a conception of human flourishing that is, in various important respects, fundamentally opposed to the reduction of human life to information. This opposition concerns not only economic, epistemic, and colonial power. The value of limits to what one can know about oneself and others is not merely a function of resistance to one type of power or another. Indeed, both Foucault and Glissant are unusually skeptical of resistance, which often seems to further solidify the reality that it purports to resist.[72] Rather, the argument against information is also an ethical one, significant for our relation to ourselves and others and therefore crucial for understanding how our practices of knowledge impinge upon or expand well-being.

As we have seen, the separation between the ethical view and the political one cannot be made so neatly, for the existence of oblivion's opportunities for self-detachment depend, like the confidence that one's life is one's own and worth living, on a certain arrangement of a society's material and ideological structure. Among these social supports are rights and practices of privacy, as well as other rights and practices against information, be they limitations over access to knowledge of our pasts or resistance to the over-documentation of human life. We do not have to think that this supports a right for individuals to affect every instance in which their lives are entered into the field of documentation in order to recognize the moral force of the idea and its relevance to our own concerns with the reach of information-making technologies.

Sleepwalking toward Insomnia

It is striking how commonly the connection is drawn between insomnia and the inability to forget. "I am memory come alive, hence my insomnia" writes Kafka.[73] Or Emil Cioran, one of history's great insomniacs: "Anything is preferable to permanent wakefulness, to that criminal absence of forgetfulness."[74] The analogy resonates because it links a pair of experiences that are quite distinct in obvious ways but alike in one: they are both experiences of insufficient access to oblivion. The story of Funes came to Borges, so he says, after many sleepless nights: "I did my best to forget myself, to forget the room I was in, to forget the garden outside the room, to forget the furniture, to forget the many facts of my own body, and I couldn't do it."[75] Borges's own inability to detach from knowledge of himself and the world around him, and Cioran's description of insomnia as a "dizzying lucidity," echo Funes's flight from his "unbearably precise" and "lucid" reality toward the dark houses of the unknown quarter.[76]

The agony, or at least the extraordinary tedium, of sleeplessness is a symptom of the insomniac's stymied desire for oblivion. If you have ever had this experience, you will know that it is one in which the world appears to the senses with an unusual degree of particularity and precision. One hears sounds differently when one cannot sleep, feels the sheets more exquisitely, but above all one becomes so mentally and physically self-aware that everything else starts to be displaced by the self. Insomnia, in Cioran's words, "kills of the multiplicity and diversity of the world, leaving you to your private obsessions."[77] The long minutes and hours of insomnia comprise what is perhaps the most common experience of wishing to detach from oneself, to let go of the integrity of personal identity and its memory, self-knowledge, self-awareness, and self-accounting for the unaccountable respite of sleep's oblivion. In a manner of speaking, it is not unlike hiding. In insomnia, I am overwhelmed by the integrity and presence of myself, my memories, and the normative commitments of personal identity: the three come together in the nocturnal replay of embarrassing moments or unkind words. As in hiding, my self—which under other circumstances is a source of value, agency, and much more—becomes a sort of sentence, and I *endure* insomnia in the same way one endures hiding, waiting to be found by sleep as if by a seeker, checking the time or willing myself not to check it just as the hider peeks out through an aperture or wills herself to abstain. The insomniac is, in Alphonso

Lingis's reading of Levinas, "held to being, held to be,"[78] and wishes to sever this tether, if only for an hour or two. The broader analogy to our discussion of the oblivion of privacy is clear.

Seventy years before Borges invited readers to think critically about the value of memory to human life, Nietzsche did the same with basically the same example:

> Imagine the most extreme example, a human being who does not possess the power to forget. . . . Such a human being would no longer believe in his own being, no longer believe in himself . . . in the end he would hardly even dare to lift a finger. All action requires forgetting, just as the existence of all organic things requires not only light, but darkness as well. A human being who wanted to experience things in a thoroughly historical manner would be like someone forced to go without sleep, or like an animal supposed to exist solely by rumination and ever repeated rumination. In other words, it is possible to live almost without memory, indeed to live happily, as the animals show us; but without forgetting, it is utterly impossible to live at all. Or, to express my theme even more simply: *There is a degree of sleeplessness, of rumination, of historical sensibility, that injures and ultimately destroys all living things, whether a human being, a people, or a culture.*[79]

We have seen several ways that such a person may no longer believe in his own being. Funes lost belief in his own life because there was hardly any of it left; he was less a human being than a mirror filled entirely with the world, with precious few obscure corners and blind spots to house a self that was his alone. His thoughts, too, could hardly be called his. Instead of rising out of and sinking back into the depths of mental oblivion, as thoughts do in normal cogitation and recollection, where they morph over time and perhaps resist our demand that they come when we call, Funes's memory is more like a collection of photographs or a computer hard drive. It is all there, unchanging and fixed, instantly retrievable. By becoming perfectly self-identical with himself at every point in his life, and by having full control over a perfectly accurate self-knowledge, Funes loses his individuality.

An excess of historicity also caused K no longer to believe in himself or his own being, albeit in a different way. He lost confidence in that central capacity of human agency to change and become different from what he was in the past. The loss of the confidence that his life was up to him (another form of *belief* in himself, indeed in his human reality) undermined his agency. Finally, we saw how the development of documentary practices gave those living through it a sense of life becoming more fixed, more factual, with less ambiguity and life-giving potentiality. This was more than a moral panic, for how a society creates, stores, and commemorates facts about the world and the past inevitably shapes its members. To make healthy and free individuals, capable of change and surprise, human beings need a balance between oblivion and information.

Late in life, once he had gone blind and lived entirely "in memory," Borges said something similar in an interview about Funes: "You should go in for a blending of the two elements, no? Memory and oblivion, and we call that imagination."[80] This attitude, from a man for whom memory offered the only connection to the great portion of the world that exists beyond our capacity to hear, touch, or smell it, reflects the normative implication of Foucauldian self-detachment and the wisdom of the paired maxims at Delphi. It is hard not to admire Borges's ability not to cling too tightly to himself and the world in order to leave sufficient room for the potentiality and play of imagination. But then again, the realm of the imagination is the domain of greatest human freedom precisely because of its qualities of flux and disintegration of personal identity. It is in the free play of the imagination that we often feel most like ourselves, at one with our human faculties and potentiality, at the same time that—and indeed because—we actively renounce the fixity of identity. This freedom, like the faculty of imagination itself, depends upon a degree of detachment from the self, from one's history and world. A society without sufficient oblivion would appear shallow and impoverished in the same way that an individual faculty of imagination would seem shallow and impoverished if it were limited to publicly recognized facts and the shared consensus on what is real and possible in human affairs. To see how this is so, let us turn to the next chapter, concerning the production of human depth.

5

PRIVACY AND THE PRODUCTION
OF HUMAN DEPTH

Hang on to sleep a little longer.

The reports from the insomniacs in chapter 4 directed our attention to the psychological and existential value of sleep in addition to the biological importance it holds for practically all animals. If more or less all creatures must sleep to live, some of them—at least those with self-understanding and a concern for personal identity—also need sleep's nightly dissolution of the self in order to live well. Yet not everyone values sleep so highly as the insomniac. Vladimir Nabokov speaks for our optimizing age of rational control when he calls it "the most moronic fraternity in the world" and "the nightly betrayal of reason, humanity, genius."[1] And although most of us are probably not quite so grandiloquent as Nabokov, it is common to regard the oblivion of sleep as a biological imposition that gets in the way of achieving our ends. We could do more if we could do with less. To sleep is also to abdicate control over one's life by disappearing into oneself and, at least as a conscious agent, to retreat from the world of collective action and objective reality, to say nothing of the surprises and enigmas that await in dreams.

"Sleep is such a dangerous place to go from consciousness," writes Jenny Diski. "Who in their right mind would give up awareness, deprive themselves of control of their senses, volunteer for paralysis, and risk all the terrible things (and worse) that could happen to a person when they're not looking. . . . As chief scientist in charge of making the world a better place, once I'd found a way of making men give birth, or at least lactate, I'd devote myself to abolishing the need for sleep."[2] This is not, as far as we know, biologically possible. It is also impossible in another sense concerning the realm of value and meaning in human life. For if we suppose for the sake of argument that Diski's dream were a possibility within reach, then we may appreciate how a world in which the oblivion of sleep was no longer a necessity of human life would be one in which a massive increase in the powers of perception, knowledge, and control comes at the cost of a significant diminishment in the depth, potential, agency, and meaning that make life worth living. So far, this book has shown how these goods rely upon the oblivion produced by privacy, forgetting, and other practices against the creation of information. Now we will see that the depth-giving goods of oblivion are at once personal and collective, private and public.

To motivate that discussion, let us consider for a moment what the world would look like if Diski had her way. Doubtless there would be benefits to life without sleep. The average person spends about a third of her life in slumber. Surely we would be able to accomplish much more with our time on earth if we never had to drift off. Gone, too, would be the deficiencies of control, reason, and self-identity that steal nightly upon sleepers. It is no coincidence that these gains in efficiency and mastery are also the sorts of benefits that are often promised to us by new technologies. Indeed, the desire for a life without sleep looks a bit like a type of hubris characteristic of the modern and postmodern age, in which the drive for more leaves us with less, and our efforts for more life result in a planet, a world of human activity, and a self that are less inhabitable, less humane. There is a world of difference between the "more" we get from a sleepless world and the "more" of the epigraph to this book that, at the end of life, the speaker of Stevens's poem "would have wanted": not a greater quantity of anything but "some true interior to which to return."[3]

A world without drifting off or waking up is one in which each human life is no longer structured around the quotidian event of emerging into the world of conscious experience, knowledge, self-control, and shared action out of unknown and uncontrollable depths that are nevertheless human depths, both integral to the individual person and shared universally among all people. From the oblivion of sleep we emerge into our lives as if onto a newly prepared stage. The sense of natality we get from the regular emergence from sleep is reflected in the perhaps overly cheery saying "it's a brand new day." Sleep has long been described as the cousin or brother of death, and to awake is almost to be reborn. But then we all fall back into the oblivion of sleep, pulling ourselves off the stage of living on an almost nightly basis. Our daily lives, our economies and family routines, everything is structured in some way or another around the regular descent into a realm of unknowing, potentiality, risk, and vulnerability. This descent is at once individual and collective, and it is unavoidable. The inability to will ourselves awake, to exercise agency in dreams, or to remember what happens in sleep are some of the most common experiences we have of confronting the limits of our ability to control our lives, to protect and to know ourselves. By the same token, the experience of emerging from sleep, when for a moment one doesn't know where one is, even who one is, only that one is alive, is another type of experience that combines the disintegration of personal identity with the confidence that one's life is one's own. When I imagine a world in which this daily experience of oblivion is neither an individual necessity nor a commonality of collective life, I see a world of terrible, flattening lucidity. Life without the experience of constantly emerging out of oblivion and falling back into to it strikes me as depressingly shallow, such that I would expect people lacking the daily oblivion of sleep to seek it, like Funes, by other means. A world without sleep is a world with one fewer region of human life that is inherently uncontrollable, full of surprise and potentiality, essentially unfathomable but nevertheless emphatically human.

I leave the exercise here, playful and at most half completed, as an indication of yet another thread by which oblivion is woven into the fabric of everyday life to powerful effect. For it is time to draw the arguments of this book together to say more about the individual and collective goods of oblivion, which we will do once again with the help of someone who is not the most obvious ally for thinking about the goods of privacy: Hannah Arendt. Arendt famously prioritized the public realm of human endeavor, where one acts and is seen among others, by contrast to which privacy often appears in her work as a kind of social death. Even when privacy is not described in *The Human Condition*'s language of political and ontological "privation," it still takes the form of something like oblivion. At times she seems to gesture toward a positive view of oblivion as a source of liveliness and possibility in human affairs. There are moments in her books and correspondence when oblivion appears not to deprive but to radiate value for human life without, however, losing its quality of essential impenetrability—a little like the light that emanates from the photon ring around a black hole.

We will begin by reconstructing an Arendtian view of the positive goods of oblivion, drawing from the discussion of privacy and oblivion developed over the course of this book. I should emphasize the reconstructive quality of my reading at the start, since "oblivion" is not a term Arendt herself ever used in this context, although we will see that it captures and expresses an insight about the public goods of privacy that lies scattered throughout her writings. However, we will also move beyond the Arendtian frame to present my own account of the individual and public goods of oblivion—including the goods of trust in oneself and others—with the goal of producing what Arendt herself, borrowing from Walter Benjamin, once described as "the deadly impact of new thoughts" that makes us see the world afresh.[4]

Arendt's Social Ontology of Privacy

It is grandiose but only barely an exaggeration to say that Hannah Arendt wrote to save the world. That I use "world" here in the somewhat technical sense particular to her work detracts almost nothing from the statement. A great deal of what she wrote, especially the books and above all *The Human Condition* (whose working title was *Amor Mundi*), was aimed against what she saw as the "world alienation" of the modern era.[5] The loss of the world or a sense of our place in it was responsible, she thought, for a stark diminution in opportunities for human flourishing and laid the ground for the acceptance of totalitarianism and its consequent horrors. In response, Arendt wrote with the hope of showing readers what they were in danger of losing—to force us, as she put it, "to think what we are doing."[6] Of highest importance in her diagnosis of world alienation was the loss of a very particular kind of public life.

We cannot talk about privacy in Arendt without talking about the public, and we cannot talk about the public without first talking about publicity. Publicity for Arendt is the condition and activity of "being seen and being heard" by others from the inexhaustible variety of their plural points of view.[7] What distinguishes Arendt's view of the public from the commonsense conception is that the epistemic condition proceeds and indeed creates the physical public space *as public*. Publicity is the ground of the public realm as a spatial or political entity, and not the other way round. "Only where things [and persons] can be seen by many in a variety of aspects without changing their identity, so that those who are gathered around them know they see sameness in utter diversity, can worldly reality truly and reliably appear."[8] This activity of perceiving and sharing perceptions by the plurality of human perspectives is what brings the public into being, not once and for all but constantly as a process of perpetual renewal. And what is true for the public is true for the world itself.

The idea that Arendt wanted to save the world by showing it anew reflects one of her most important and frequent claims: that the world and reality as we know it are human achievements. This isn't exactly a new idea, but Arendt's version of it, which gives socio-ontological pride of place to the condition of publicity, is of novel force and insight. Arendt draws a distinction between the earth, which is the planet from which all living things draw sustenance, and the world, which is an artifice of human life and civilization built on and out of the earth. The world consists of the material objects we make and whose public existence exceeds the span of individual lives— buildings, paintings, monuments, tables, books—and whose reality depends, like all worldly "objective" reality, not on their physical properties but on their appearing in the light of publicity. "The whole factual world of human affairs depends for its reality and its continued existence, first, upon the presence of others who have seen and heard and will remember, and second, on the transformation of the intangible into the tangibility of things."[9] What is not thus transformed cannot be perceived and known by the human plurality, which means that it will never attain the quality of *really existing* that is bestowed by the light of publicity.

What is true for the reality of worldly objects is doubly true for human beings, who are not quite so durable and long-lived as a table and who are known in a different way than such objects. We can ask "What is that?" about both people and tables, but only people are fitting objects of the question "*Who?*" Therefore, human beings only gain reality by entering into the public "space of appearance," where they can be seen and heard appearing and acting in ways that mark them out as one particular individual or other.[10] This intangible space of appearance is like a stage, for which the material world provides the proscenium, set, and props, and upon which human beings gain their reality as individuals by acting and speaking in concert with others who see and hear them. Against the modern notion that the truth about oneself is something discovered in private or in intimate relations, Arendt argues for a view of individuality that is gained only in the light of publicity and human interaction. This was and remains an untimely idea about the self and its reality, so to help us understand, Arendt dramatizes her argument with a stylized, quasi-metaphorical depiction of the ancient Greek and Roman polis, where the "public realm . . . was reserved for individuality; it was the only place where men could show who they really and inexchangeably were."[11] This philosophical pseudo-history is not meant to offer an exemplar of an ideal politics to which we should return, but rather is an estranging lens meant to shock us into seeing our own world with fresh eyes.

If the light of publicity is the life source of reality in Arendt's symbolic polis, the oblivion of the private imposed a condition of nonexistence on the women and slaves confined to it. As Arendt puts it in one of several such passages: "In ancient feeling the privative trait of privacy, indicated in the word itself, was all-important; it meant literally a state of being deprived of something, and even of the highest and most human of man's capacities. A man who lived only a private life, who like the slave was not permitted to enter the public realm, was not fully human."[12] Passages such as this one make it hard to imagine there could be any benefit to such oblivion. It resonates, too, with feminist critiques of the use of privacy as a tool of patriarchal domination and cover for all sorts of horrors and abuses. One imagines that battered wives and abused children felt something like what today we call gaslighting, owing to the public's fastidious ignorance of their suffering, expressed in everything from rape-shield laws to neighbors pretending not to see; it must have given victims a sense that, for the world at large, it was almost as if didn't happen. But Arendt suggests an even stronger critique: that women confined to the home were denied a public existence, where they would be seen and heard by a plurality of others, and in that sense it was as if they didn't exist at all, at least not as men and other objects of public attention did. Arendt obviously thinks this is a terrible harm to suffer, yet she does not therefore conclude that privacy should be done away with, or that its condition of total oblivion be exchanged for mere opacity. Rather, she defends the oblivion of privacy because of the positive goods that it lends to human life, what she called the "non-privative characteristics of privacy."[13]

Arendt thought there were two ways that oblivion did not deprive from but in fact added to human life: one having to do with those aspects of experience that cannot by their nature appear in public, the other with how oblivion's "darker ground" lends depth to the life of both individuals and the public realm they inhabit.[14] Both views of oblivion are underdeveloped in Arendt's thought, which of course is no surprise, given her intense focus on the public dimensions of human life. But they are also so rich that they suggest others.

We Do in the Dark What Wilts in the Light

Arendt thinks that there are certain elements of human existence that can survive only in private. This is not an argument that some things are shameful and therefore should be kept from public view.[15] Rather, the idea is that "a great many things cannot withstand the implacable, bright light of the constant presence of others on the public scene."[16] At times, Arendt sounds a direct echo of nineteenth-century privacy advocates—for instance, when she supposes that it is on account of growing up in a condition of publicity, even in the "private" home, that "the children of famous parents so often turn out badly."

> Fame penetrates the four walls, invades their private space, bringing with it, especially in present-day conditions, the merciless glare of the public realm, which floods everything in the private lives of those concerned, so that the children no longer have a place of security where they can grow.[17]

Glare nicely captures the sense in which the metaphorical floodlight of publicity can be harsh and overwhelming, but also how this light is no light at all but the public attention of others. And her hypothesis about the children of celebrities recalls Warren and Brandeis's complaint of sixty years prior that "no enthusiasm can flourish, no generous impulse can survive under [publicity's] blighting influence" and the worry expressed in "The Sacred Privacy of Home" that a deficiency of privacy posed an impediment to an individual's "social and moral cultivation." This strain of argument in Arendt has been interpreted by insightful readers as advocating privacy as respite from the demandingness of life as it is lived in public. Seyla Benhabib, for example, tends to understand that demandingness either through the lens of Arendt's call for a vigorous civic-republican public sphere of agonal politics, according to which privacy "protects, nurtures, and makes the individual fit to appear in the public realm"; or in a liberal sense of the private as a realm for the development of personality, according to which privacy "provides the self with a center, with a shelter, with a place in which to unfold capacities, dreams, and memories, to nurture the wounds of the ego."[18] This is all sensible enough, although I will leave aside the civic form of respite and suggest in its place the agential, epistemic form described in chapter 2, since this would better explain why we should need to rest from the exigencies of mere *appearance*, independent of political action. Indeed, one can rest from political activity in public, say by chatting about love with friends in an olive grove, sitting silently at a drinking party, or simply going for a walk. Benhabib's second hypothesis, however, is quite far from Arendt's own view. A conception of privacy as a place to discover one's authentic self would have been anathema to the author of *The Human Condition* who, in the preface to that book, singles out "the flight . . . from the world and into the self" as the principal cause of the world alienation she seeks to understand and perhaps remedy.[19]

Arendt claimed that privacy protected certain valuable experiences that could not withstand "the glare of publicity" or the "public aspect of the world" and its quality of objective reality. Among these valuable things supposedly was love, "which is killed, or rather extinguished, the moment it is displayed in public. ('Never seek to tell thy love/ Love that never told can be.')."[20] This seems obviously wrong to me—we *do* love in public, all the time, in both amorous speech and action. However, the pair of lines quoted from William Blake, and the context in which they appear, suggest that she meant something else by this. Blake's poem is a mischievous piece of didactic verse about someone whose confession of love is unrequited with the result that the beloved runs off and the friendship is ruined. By contrast, Arendt's strong misreading centers around there being something essential about love that resists being put into words.[21] Getting "told" and "being" are mutually opposed; the former cancels out the latter. If an emotion or experience consists in something that is "extinguished" when put into language—that is, when it undergoes the "transformation of the intangible into the tangibility of things"—then it cannot survive the condition of publicity, which would alter its nature irreversibly. The idea that there are in fact many such valuable elements of human life which have this quality of innate resistance to being articulated is one of the repeated arguments of this book. It is also an underappreciated element in Arendt's thought.

In her later work on the inner life, Arendt distinguishes between two modes of self-relation. There are "the operations of the mind," which include thinking, language, concepts, and information. And then there is "the life of the soul," by which she means those forms of self-relation that do not natural-ly take the form of propositional knowledge: myriad "feelings, passions, and emotions," prominent among which, of course, is love.[22] Whereas thoughts and ideas take form in language—"Thought without speech is inconceiv-able"[23]—the vast range of human experience classed under "soul" does not naturally take the form of discourse and is therefore deformed when put into words. Actually, she thinks that these experiences cannot survive the transi-tion into language: they "can no more become part and parcel of the world of appearances than can our inner organs."[24] It isn't that these parts of human life are shameful or embarrassing, but are "killed, or rather extinguished" by the condition of publicity—that is, when they are made to assume a form that can be spoken, known, and communicated in the impersonal form of information.

The constitutive inarticulacy of this range of experience and self-relation means that it needs the oblivion of the private to exist at all. This explains why Arendt would call "great bodily pain the most private and least com-municable of all" feelings and experiences—it is private because it cannot be communicated, only summarized or evoked in language.[25] Thus she uses the lines of Blake not, as Blake himself does, to recommend keeping your crushes to yourself lest you lose a friend, but in support of the following line of argu-ment. There are aspects of human life that constitutionally resist knowledge and communication, which therefore are harmed or dissolved when turned into a form that can be known and readily communicated in literal speech (like data and information). These aspects of our lives are part of what make them worth living and therefore deserve our protection against potential threats, among which is their translation into information. The condition of publicity entails the production of its objects as knowable: it is the condition under which the "intangible becomes the tangible." Therefore, these valuable, unarticulated domains of human life give reason to maintain the protections of privacy against the encroachment of publicity.[26] The "sacredness of this privacy . . . harbors the things . . . impenetrable to human knowledge" not because we cannot know about them (she has in mind here stuff we all go through: birth, death, sex, yearning, love, pain).[27] Rather, they are impene-trable to human knowledge in a more general sense *because* we shield them off from public view. The social practices of privacy do not just respect the naturally occurring impenetrability of these deeper realms of human life:

they directly contribute to the sense of their being impenetrable in the first place. In other words, the protections of privacy screen off or *set aside* certain realms of experience as beyond articulation and knowledge in the literal sense of the "sacred."

In light of chapter 1's discussion of "sacred privacy," we should not be surprised to find that Arendt herself frequently refers to privacy or the private realm as "sacred."[28] We don't need to agree with Arendt about what sorts of self-relation, knowledge, and experience properly belong to this realm. In fact, it is probably best to leave it open, both as a conceptual matter and as a political question about the discovery of new and individual modes of flourishing apart from, against, and beyond information. We need not think that love fits there, although I tend to think it does, at least on Arendt's terms. Recall that love cannot withstand publicity not because it is shameful, but because the condition of publicity turns its "intangibility into the tangibility of things" and thereby diminishes love's essential and meaning-giving inarticulacy. Think of how love is notoriously difficult to "put into words" in the sense of satisfyingly exhausting it with meaning. It is by reference to this idea that we are made to see Cordelia as the only one of her sisters who actually loves their father when, after her sisters put their affection into words, she says: "Unhappy that I am, I cannot heave my heart into my mouth."[29] The experience of loving someone or being loved is so potent—and love seems to offer such a vital source of meaning, power, and depth in our lives—in part because it "passes understanding" and outstrips our efforts to understand, explain, and communicate it without remainder *while nevertheless remaining integrated into our lives* in the most intimate and sustaining way.

This helps us to understand of one strain of contemporary thinking about privacy, which finds its moral basis in the protection of intimacy.[30] On its face, the notion that intimacy is impossible without privacy seems dubious: anyone who has ever spent time in a city where teenagers are forced to seek public arenas for their canoodling knows that it isn't literally true that privacy is necessary for us to engage in intimate activity without inhibition, embarrassment, or shame.[31] Indeed, what we envy when we see lovers displaying affection for one another in public is that for them, the world has fallen away. They are *oblivious* of their surroundings, the leering or sarcastic glares of onlookers, the oncoming bus, all of it. It is oblivion that protects intimacy, which privacy reliably produces but, as the lovers demonstrate, can be achieved by other means, as well. Without oblivion of some sort, one isn't "really present" in the activity. That the necking lovers create something similar to privacy—namely, oblivion—gives credence to the Arendtian claim that some forms of experience are constitutionally opposed to being converted into language or information. Privacy isn't necessary for this, but oblivion is.

If I approach a pair of lovers kissing in Central Park and take a photograph of them, I have acted badly. I have wronged them and perhaps even harmed them. The same might be said, albeit with a lesser degree of seriousness, if I tap them on the shoulders until they stop and look at me. The most obvious way of describing my act is that I have violated their right to be let alone—which, of course, is one of the more famous descriptions of the right to privacy.[32] I should mind my own business and keep my nose out of theirs. But of course we don't have a general right to be left alone in a public park, nor do we have a right to privacy there. In this case, the lovers never had privacy for me to violate, and it stretches the ordinary meaning of the word too far to contend that they did. Rather, what I have done is I have destroyed or altered their oblivion. At the very least I have acted badly by disrespecting their right to access certain forms of oblivion even in public. To be sure, when describing such a scenario, we tend to reach for the language of privacy, notwithstanding the obvious fact that we use such language by analogy and not as a literal description. We do this because privacy is the most common and widespread way that we protect and produce oblivion. But as the lovers and many of our other examples show, privacy is not necessary to this end. It just turns out to be the most obvious and reliable of several ways that oblivion is produced in society.

The Darker Ground

Here things get exciting. The obscuring of those parts of life that cannot withstand the light of publicity is neither Arendt's most important nor her most mysterious suggestion concerning the value of oblivion. In what might be the most poetic and least understood passage of *The Human Condition*, she gives a second view of oblivion's generative power for both individuals and the public realm itself.

> A life spent entirely in public, in the presence of others, becomes, as
> we would say, shallow. While it retains its visibility, it loses the qual-
> ity of rising into sight from some darker ground which must remain

hidden if it is not to lose its depth in a very real, non-subjective sense. The only efficient way to guarantee the darkness of what needs to be hidden against the light of publicity is private property, a privately owned place to hide in.[33]

What puzzles and moves the spirit is the image of rising up from "some darker ground" whose "depth" consists in oblivion—a funny kind of depth, since one cannot actually see into it (a necessary condition of perceptual depth). The depth of oblivion is a depthless depth, or better: the intimation of a beyond that does not reveal it. But what most puzzles and moves is the idea that without this epistemically impenetrable depth, human life would be "shallow"—that is, lacking in depth of another kind having to do with the meaning and quality of a life. Let us call this second sort *human depth*, in recognition that it has something to do with a normative picture of human well-being. Arendt does not explain just how such a life would be shallow, though she does seem to think we have a common sense about it. We probably have some intuitions in the area, but we had better start with its opposing term of depth, about which Arendt is more specific.

The epistemic depth of oblivion, which is "hidden from human eyes and impenetrable to human knowledge,"[34] is like a pitch-black night, which is better described as *depthless*, since depth, in its ordinary meaning, means something that can be penetrated—by the eye in the visual field, a plumb in a body of water. In English, we might call this type of depth *unfathomable*, a word that describes its object as both very deep and impossible to measure. The footnote to the quoted passage reads: "The Greek and Latin words for the interior of the house, *megaron* and *atrium*, have a strong connotation of darkness and blackness," which recalls the life-sustaining "black, compact [houses], made of homogenous shadow" in Funes's imagination. This sort of depth is an intimation of the beyond, which, in positive terms, is a manifestation in the material and social world of a particular realm beyond the penetrative powers of perception and knowledge and therefore a testament to the existence of such beyondness in human life. The shielding off of this beyond does not just protect it, but also creates or sustains it. Thus Arendt remarks that, from the perspective of the public—that is what one rises into from the depths of oblivion—what is important is not "the interior of this realm, which remains hidden and of no public significance, but its exterior appearance."[35] What is essential is that the darker ground cannot be seen into but also, as with Funes's dark barrio, that there be some sign testifying to the non-imaginary existence of its impenetrable beyond.

Arendt's passage compels us to recall one of this book's central claims: that oblivion is a human construction, whether in perception and memory, space and time, self-knowledge and the knowledge of others, or the arrangement of the material world and the social ontology of human reality. Still, the connection of the depthless "darker ground" to the idea that a life or life in general has depth in the human, ethical sense still needs explaining, although to do so we will have to depart from Arendt. There are at least two ways of understanding this connection, which correspond to two types of good: the individual good of depth in one's relation to oneself and others, and the public or common good of depth in a society. Let us take these in turn.

Personal Depth

The contrast between a deep life and a shallow one is commonly used if infrequently made explicit. Philosophers, stoners, and everyone in between understand what we mean when we say "he's deep" or "that's deep." It is perhaps the most common, nontechnical statement about the meaning of a life or something in it. However, and doubtless as a consequence of the depth of this depth, whatever it is, the phrase is not nearly as easily defined as it is used. Let us begin by noting the obvious. To call a person deep usually means that there is more to her than meets the eye. It is not only persons who can be deep in this sense, but the term *is* restricted to the humane or what Arendt would call "the world." A statement, a painting, a film, and a poem can all be deep in this sense, but one's pockets or a body of water cannot. If we are standing on the shores of Crater Lake in Oregon, whose bottom lies nearly two thousand feet below its surface, and our friend says, "man, that's deep" in the way one would say it about a person or a koan, we would laugh. It is funny because he has conflated one sort of depth with the other, human or ethical sort. The difference between the two is not merely linguistic but expresses an opposition concerning the existence of limits to what can be known. We must know where the bottom of the lake lies to call it deep rather than bottomless. By contrast, the depth of a deep statement tends to evaporate or diminish when someone offers a sufficient explanation for the phenomena in question: the koan turns out to be a mere riddle, the poem exhausted by interpretation.

A life cannot be described as deep in the human sense if it lacks some essential quality of resistance to exhaustive knowledge. Just as a figure without a background will appear flat because it is unimplicated in a broader world, a self that does not appear against a background of the unknowable will seem to be lacking the depth of personality which, in another context, corresponds to the difference between "flat characters" and "round" or "deep" ones.[36] Flat characters in the narrative arts are those who give no sense of existing beyond the limits of the scenes in which they appear; they seem artificial because the author has not convinced us that they have a life of their own. "Round" or "deep" characters, by contrast, strike us as more real, more alive, because their lives seem to go on even when we are not looking or cannot see. They seem to have lives of their own because the author has used her craft to create the illusion of oblivion. Hence the most useful tool a novelist has to create such depth is not exhaustive exposition of every moment in a character's life, but the deliberate omission of information and the imperfectly omniscient narrator. Only round characters are "capable of surprising in a convincing way," as the novelist E.M. Forster put it, because unlike flat characters they give the impression of containing multitudes beyond what the reader (or narrator) can see and know.[37] By the same token, flat characters are flat, and the Arendtian "life spent entirely in public" is shallow, because what you see is not only what you get, but it's all there is. The difference between the two types of literary character is not just a bit of technical craft knowledge for fiction writers; it is also another way of talking about what makes a life worth living—for who would choose to be a flat character over a round one?

For example, try to bring into your mind the self-image that you present in a certain context, the persona you project to your colleagues, or your students, or your in-laws. Now imagine meeting this person. Imagine spending a day together with only the "work self" version of you and not the more capacious person who contains that and other personae, can shift between them, and who importantly exceeds the sum of all public faces. This "version person," I suspect, would appear flat, one-dimensional, and, I think, especially constrained or unfree.[38] He or she is flat—"shallow"—because what you see is not only what you get, but it's all there is. The same is true of our relationships with friends and lovers. I think of my wife. If there is anyone about whom I can say that I know them better than they know themselves, it is she. And yet, if there were no limits to my knowledge of her, even in theory, then there would be no more to her than what I could know. Where there is now an independent, autonomous person capable of surprising us both there would be something flat and unlively, more like an embodied filing cabinet than the woman I love and yearn to know as much as possible. She would be to me as Funes is to himself, with the illusory flat depth of a mirror.

If this is true about my relationship with my wife, then it must also apply to my relationship with myself. Just try to imagine what it would be like to live with yourself under the conditions of perfect and total self-knowledge. To be completely transparent to oneself, with no forgetting or blind spots: this is Funes's untenable condition. It is hard to imagine, of course, but we can try. What would solitude be like if there were no limits to self-knowledge? More like flipping through a dictionary or a photo album than the fluid, uncertain, and lively play of thoughts that come out of nowhere and sink back into it. What resources would we have for coping with boredom, failure, or loneliness under conditions of perfect self-knowledge? What would introspection be like if I were entirely accessible to myself? I may look inside to retrieve a bit of information (and with what greater efficiency!) but I would not have the experience of really *looking* or *searching* within, of coming to grips with myself, of trying and failing to recollect some image or word, of failing to compass the entirety of my knowledge and memory. Such failures of introspection may be deficiencies of agency or self-knowledge, but they are also vitally important for the sense that our lives have *depth* and contain more than what can be controlled or called up at an instant, whether by oneself or by others. (Recall the doctor's fearful scope.) The experience of *searching* within is necessary for the confidence that one does, in fact, contain searchable spaces or depths within oneself.

Arendt herself seemed to have this sort of experience in mind when she cites the Eleusinian Mysteries as a proto-manifestation of privacy's expressive and productive value: "It seems as though the Eleusinian Mysteries provided for a common and quasi-public experience of this whole realm [i.e., the realm of privacy and oblivion], which, because of its very nature and even though it was common to all, needed to be hidden, kept secret from the public realm: Everybody could participate in them, but nobody was permitted to talk about them. The mysteries concerned the unspeakable, and experiences beyond speech."[39] These mysteries aren't secrets but private or proto-private (unaccountable, even), for they are maintained *beyond* speech in the same way that privacy produces a realm of *the beyond* concerning the knowledge of individuals. Without some sort of public sign testifying to its existence, the unspeakable can have no shared, objective, worldly reality, as mysteries do for members of a particular religious community, and as the impenetrable oblivion of privacy does for Arendt's polis. This is, in a way, the point of the famous ending of Wittgenstein's *Tractatus*. By contrast, the unspeakable quality of the Eleusinian rites, which in fact were known and experienced by initiates, depends upon the social prohibition against speaking them. Otherwise the unspeakable becomes the variously described. The Eleusinian prohibition erects a discursive boundary similar to that of privacy: it doesn't protect an already existing and metaphysically essentialist realm of ineffability "beyond speech" but rather produces it by privacy's double action of concealing something by means of a barrier that *is* visible, publicly legible, and respected by the human plurality of the shared world. (The difference here between privacy and secrecy, hiding, or disappearance is highly illustrative.)

Before you object that perfect knowledge of myself or my wife is impossible, perhaps because there are certain limits to cognition inherent to the physiological processes of thought and memory, I hasten to say two things. First is that even if there were such limits to what we can know about people, the undesirability of perfect knowledge nevertheless indicates a value in human life that pits some things that are important to our lives going well—depth, oblivion, surprise, potentiality—against other important values concerned with knowledge, information, autonomy, control, and the like. The problem does not disappear at the level of value even if it does at the level of biology, which means that the question of value persists as a problem for politics and ethics, notwithstanding the impossibility of anyone turning out like Ireneo Funes.

The objection is also spurious. The claim that there are natural limits to how exhaustively we can know someone assumes the conclusion it is meant to prove: that human beings are the kinds of things that cannot be completely known by the biological processes of cognition such as we understand them. The objection depends on a certain philosophical-anthropological conception of the human being and what can be known about it. This conception of the person is not a natural fact about us in the same way that the chemical processes of neurotransmission are, but is rather an artifact of human construction. Like any other idea, the idea of the self has a history: it changes and will continue to change. If it is part of human nature, it is not a part of our biological nature the same way that sweating is, but rather a part of what Arendt might call our "worldly" nature, along with other things like dignity, the potential for change, human rights, interests, interior monologues, and so on. We can easily imagine a society of people working and living under the assumption that a person *could* be known all the way down, or even that what one knows about one's lover, say, is all there is to know. The history of our species abounds with stranger beliefs. Indeed, we need look no further than the idea in certain sects of Protestant Christianity that God knew all one's thoughts, one's future, one's acts, and even all the hairs on one's head, such that there is "nothing concealed that will not be known."[40]

More to the point, we have seen how human subjectivity has altered in response to the introduction of technologies that interrupt and mediate practices of coming to know oneself and the world. The invention of the photograph was one striking example, as was the rise of the Internet's global archive. Foucault, at the end of his life, was charting how the invention of practices of autobiographical writing gave rise to two quite different and far-reaching cultural understandings of what could be known about a person, and how.[41] The explosion of literacy and printed material in the early modern period changed our relation to history; the introduction of the mirror into the homes of average folk altered practices of self-regard.[42] The locomotive changed our perception of space, time, speed, freedom, and risk.[43] We could also mention the census, electric light, passports, statistics, and the techne of behavioral sciences—the list goes on and, of course, includes the striking developments of the digital age, notably social media, the self-tracking processes of "the quantified self," and the omnipresent surveillance of the "Internet of things."[44] Whether such change is good, bad, or indifferent is a normative question, and perhaps ultimately a political one, which must in any case be answered by reference to some picture of human flourishing. However, changes in subjectivity bring with them changes in what it means for a person's life to go well, which is why it is especially important to think about what we are doing as we are undergo such changes. Otherwise we may stand to lose a moral vocabulary for understanding the person that we would rather keep, as well as the will and intuitions that activate it in the sphere of practical ethics and politics. The problem is especially acute when it comes to the question of privacy in an age of information. To the extent that we come to think of human beings as repositories, producers, or somehow consisting of information—that "you are your data," as sociologist Deborah Lupton puts it[45]—we risk replacing a view of human depth and flourishing with an acceptance of shallow fixity and the ready manipulation of information.

The Public Goods of Oblivion

Do the personal goods of oblivion scale up to common or collective goods? Arendt seemed to think that they did, writing that "privacy was like the other, the dark and hidden side of the public realm."[46] I will leave aside her main argument about the public benefits of privacy's oblivion, which is that it keeps certain emotions, experiences, and sights out of the public realm because they are thought to be deleterious of her preferred conception of the agonal political life. This a negative sort of good whose value has to do with the absence of certain things from the public and not with the value of privacy per se. From the discussion of the depths of oblivion, let us recall a pair of curious claims. First is the one we have just been discussing: that the value of privacy's oblivion lies in its protecting and producing aspects of the human person that are impenetrable to human knowledge. The second comes immediately after and seems to contradict the first: "Not the interior of this realm, which remains hidden and of no public significance, but its exterior appearance is important for the city as well."[47]

Whereas the interior of the private realm provides shelter for individuals, it is of no significance to public, collective reality, except for the sight of the walls that shield it from view. The facades of private homes in Arendt's metaphorical polis participate in the social reproduction of oblivion by a twofold action. First, as a general matter, they mark off a region of the human world as impenetrable to perception and knowledge. Second, by presenting a visible barrier, they testify to the existence of those regions while maintaining them in their essential quality of oblivion. This two-fold production of oblivion's "beyondness" in human life was also seen in Funes's unknown houses that were "black, compact, made of homogenous shadow," and in Proust's description of Odette's windows, which, seen from the street below, reveal (by concealing) "the mystery of the human presence which those lighted windows at once revealed and screened from sight."[48]

By containing oblivion within itself—that is to say, by containing regions that are impenetrable to knowledge, self-determination, and control, but which are nevertheless *human* regions (as opposed to the cold far reaches of outer space)—the shared world of public life is deepened in the same way as a literary character. Recall what we described as the public good of the right to be forgotten: that by its very availability, the right holds open a sense of possibility in human affairs and reinforces collective confidence in the ability of individuals to be different from how they were, are, and will be. In a similar way, the idea that there is always more than meets the eye in human life lends the public square a quality of excitement, possibility, and depth. This appears in Rothman's use of Virginia Woolf's enchantment at mixing among strangers to evoke what he calls the artist's sense of privacy: "The feeling of solitude-on-display that the sidewalk encourages. . . . She was drawn to the figure of the hostess: the woman-to-be-looked-at . . . who grows only more mysterious with her visibility."[49]

Just as the facade of the private home provided material support for the existence of oblivion in Arendt's polis, Woolf's particular view of privacy lends depth to the common world and her interactions with strangers. To see how this is, you can imagine that rather than thinking that what lies behind appearances is fundamentally unknowable, Woolf operated under the assumption that behind them lay nothing (as in Descartes' famous skepticism about automata), or instead that what she did not have access to were the secrets of others, their hidden lives. The point here is one we've made several times: that how we understand barriers to knowledge, and why they are valuable, has far-reaching consequences for human life in both private and public.

The oblivion of privacy also lends a sense of meaningfulness and possibility to the public world of politics and identity formation. If, as Arendt thought, the quality of publicity was like being on stage—"Living beings *make their appearance* like actors on a stage set for them"[50]—life in public needs something that the theater does as well: a sharp distinction between the bright light of the stage and the oblivion beyond its borders. A stage play requires a particular kind of suspension of disbelief related to the frame of the proscenium (taken here as a metaphor—this applies equally to street theater and Arendt's public). When an actor steps on the stage, the audience and the other actors actively forget or ignore whatever they may know about the actor's personal life, the other roles they have seen her play, and so on. The actors' job is to convince the audience that they are who they say they are by recourse only to how they appear, what they say and do, within the frame of the proscenium and duration of the play. This is what I meant when I said that oblivion lends public life a degree of *meaningfulness*: the reasonable confidence that one's actions on stage are certain to *mean* something lends a degree of worth to their undertaking, which in turn contributes to the sense that one's life *means something* as well (in other words, that it is worth living).[51] This is another way of putting K's complaint about the right to be forgotten: the sense that all his interactions will be prefaced by knowledge of his crimes leads him to doubt whether his actions, his life, will ever have this quality of meaningfulness again.

The public plays an equally important role here by remaining oblivious to everything except for what transpires onstage. Through their cooperation, a particular type of public space is born—what Arendt would call "the space of appearance." Just as the stage requires a frame, both material and epistemic, this sort of public needs privacy's oblivion for its quality of public reality: "Our feeling for reality depends utterly upon appearance and therefore upon the existence of a public realm into which things can appear out of the darkness of sheltered existence."[52] In political terms, Arendt's public realm is most typically described as agonal because its vying for identity and collective action is not structured as a scripted play is. But as far as the ontology of the person is concerned, it is better described as dramatic. That oblivion not only makes this possible but lends depth and plausibility (in Arendt's terms: reality) to appearances is reflected by analogy with the experience of going to see a famous actor or a relative perform in a stage play—Jim Varney as Hamlet, say. The inability to be oblivious to what one knows about the actor robs the character on stage of some of its reality and therefore its depth as a mimesis of an actual human being. In just the same way, the Arendtian public world of objective reality and agonal politics requires a background of oblivion for the collective activity of worldmaking and the individual pursuit of identity.[53]

Oblivious Trust and Trustworthiness

I want to focus on another valuable element of human life that depends upon limits to what can be known about persons: trust in others and oneself. The idea of self-trust is a relatively niche topic in epistemology and ethics, but it offers another important angle on our discussion of oblivion. We can bring this into sight with an experience that I expect everyone has had at some point in their lives.

Think of a time you had a difficult problem for which, despite your best efforts, you could not come up with a solution. Maybe it was a tricky line in a poem you were writing or an engine that refused to turn over, a step in an argument or a dripping faucet. Sometime later, maybe while you are engaged in another task or in the middle of the night, maybe while you are still wrestling with the conundrum, a solution comes to you, as the saying goes, "out of nowhere." When we say that an idea comes to us out of nowhere or "out of the blue," we do not mean that we once knew it but simply could not recall it at will—this is also a common, but quite different, experience.[54] Nor do we mean that we arrived at the solution by the deliberate combination of available materials, as in answering a complicated problem in calculus or like Sherlock Holmes induces whodunit from the available evidence. Rather, the experience is like an artist's inspiration: "Like a bolt from beyond, it occurred to me." The common use of the passive voice and metaphors of externality to capture this experience reflect the conviction that the idea, which came from the brain inside my skull (no one actually whispered it into my ear or stuck me with it like a beer can hurled from the window of a passing car) nevertheless came from a part of myself that is *beyond* me. I contain regions that are beyond me but are nevertheless part of me in the same way that the public sphere contains and draws depth from the oblivion that privacy produces by shielding it from view. The "nowhere" in "out of nowhere" is not a spatial metaphor but a metaphysical one: out of no-thing (oblivion), a thing appears. Yet the "nowhere" or "nothing" of oblivion, out of which the idea emerges, is not literally nonexistent; we can contact and experience it, although we cannot reduce it to knowledge or perceive its limits. We call these inner realms "depths" or characterize them as dark precisely because they resist the penetrating gaze of introspection and, because we understand their impenetrability to be that of oblivion, they seem to go on beyond the boundary at which even our mightiest powers of introspection and self-knowledge stop.

When we *come up with* an unexpected solution or an idea, it is more like having dived into our internal depths of oblivion than retrieving it from a cupboard or filing cabinet. Indeed, the experience often feels like surfacing from a deep dive into the unarticulated regions of the self and emerging breathless and excited into the light, the visible, knowledge, and the world of others—from the realm of potentiality to the realm of Arendtian reality—as if holding the new idea aloft to show it to others or to see it better ourselves. Thus the urge to share the brilliant thing we came up with, as if to confirm its reality in the eyes of another. It doesn't always happen like this, of course. Sometimes *we come up empty handed*; to do so over and over is as exhausting and frustrating as diving repeatedly into an actual sea only to resurface time and again without having retrieved the desired object. In any case, this is one of the most common forms we have of making practical contact with the notion of human potentiality, which is why the experience can feel so thrilling even when the solution concerns something as humdrum as a dripping faucet.

Now that we have seen how this form of internal oblivion and our experiences of acquaintance with it—that is, not "out of my memory archive" but "out of *nowhere*"—contribute in a general way to the sense that one's life is deep rather than shallow, we can get a more specific view by considering how it also conduces to trusting oneself.

There are two ways of understanding the kind of self-trust that comes from this experience. One is evidentiary and corresponds to the prevalent view of other-directed trust as a belief or rational attitude.[55] Since I've had this experience before, the next time I need an idea or the solution to a problem, I'll be patient and trust that one will come. Why? Because it has come before. But this sort of self-trust is defeasible. If I have a string of failures—say, in sum (three failures to one success) or in recent times (good ideas used to come to me but no longer)—then it would be rational to lose trust in myself on this view. Maybe this is good in some cases, like the alcoholic not trusting himself to have just one drink. But in any event, this kind of self-trust is not a feature of one's depths of oblivion, but of self-knowledge and control, in the case of the alcoholic, or memory, in the case of one who trusts that he can recall all fifty state capitals because he has consistently done so in the past.

When it comes to ideas "out of nowhere," a loss of self-trust seems to indicate a more serious harm to the conditions of agency and the quantity and quality of depth in a person's life. This is because there is an even more fundamental form of self-trust than the evidentiary one, which is directly connected to the sense that one's ideas rise up from an internal darker ground that is a part of oneself but nevertheless beyond one's powers to penetrate with knowledge and control. This type of self-trust derives from the lived, if not always self-aware, confidence that who I am does not end at the limits of what anyone can know, remember, or access, but rather contains regions beyond them that can never be wholly known or accessed. Emerson anticipated this connection between self-trust and unfathomability: "For this self-trust, the reason is deeper than can be fathomed—darker than can be enlightened."[56] Think back to how I introduced the example. The metaphor "come up with" already pointed to this connection of personal oblivion and self-trust: it is as if one dives into the murky depths of oneself in hopes of coming up with a surprising idea or sparkling rarity, though just as often one resurfaces with empty hands and out of breath. If there is more to me than meets the eye, even the eye of introspection, then I have reason to believe that if I am ever called upon to respond to a situation for which I have no solution or ready response, indeed something that exceeds all my known capacities and frameworks for understanding, I will have unknown depths to call upon. Of course, I might fail in such a situation. I often do. But these depths are also useful in dealing with such failure, partly because the basis of this stronger form of self-trust in oblivion renders it somewhat immune to being undermined by evidence. Actually, it would seem that we need to fail sometimes. The experience of coming up empty handed contributes to the sense that we have depths in principle unknowable to us, which contain much that cannot be called up at will. This gives human life a quality essentially resistant to instrumentalization, for what cannot be exhaustively known, recalled, or predicted cannot be entirely controlled. Likewise, if I ever come to dislike the way I am or have been, the idea that there are parts of me that are perpetually undefined supports the sense that I can take my life in a new direction.

It is obvious how this type of self-trust forms a valuable mode of self-relation that has its basis not just in limits to what can be known about ourselves (all trust requires imperfect perception and knowledge) but also in the idea that beyond these limits to self-knowledge lies not void or secrets, but oblivion. The relation of self-trust to self-esteem and self-respect also seems clear: people who cannot trust themselves would likely suffer the same kind of loss of self-respect and esteem that they would if other people refused to trust them, and vice versa. The deeper form of self-trust also plays an important role in agency and the self-determination of one's life. For one thing, it seems to be another aspect of an agent's sense that his life is up to him to direct and worth the trouble it takes to do so. Recall Charles Taylor's idea that self-directed agency requires the ability to come apart from oneself and reevaluate one's normative commitments. Taylor, whose theory also associates human depth with the unstructured and inchoate, argues that the "ordinary metaphor of depth applied to people" reflects a positive judgment about those who undertake this act of self-reflection, but that it is most fitting to call deep those who are capable of disassembling the self entirely in a "stance of openness" and in contact with their "deepest, unstructured sense of things."[57] In chapter 2 we argued that privacy is necessary for this experience; self-trust, too, would seem to be vital for it, since openness to being changed by "the chartless darkness of the human heart" is a risky endeavor indeed, in which the self is fundamentally vulnerable.[58]

The relationship between self-trust and the coming apart of the self goes both ways, however, for the ability to come apart and encounter unknown regions of oneself is also a resource for self-trust. Again, in his essay on self-reliance, Emerson seemed to have had a similar connection in mind between self-trust and the coming apart of self-identity: "The other terror that scares us from self-trust is our consistency; a reverence for our past act or word because the eyes of others have no other data for computing our orbit than our past acts, and we are loth to disappoint them."[59] Note that Emerson has in mind what I have called the second, deeper form of self-trust, which is not based on evidence from one's past but is in fact opposed to such evidence. Indeed, consistency is a terror that scares us from self-trust. Yet even this deeper form of self-trust must be grounded in *something* for it not to be a delusion.

Emerson's answer to this challenge is quite like mine, albeit phrased in the idiom of his day. Emerson asks what is the object of self-trust. "Who is the Trustee? What is the aboriginal Self, in which a universal reliance may be grounded? . . . The inquiry leads us to that source, at once the essence of genius, of virtue, and of life, which we call Spontaneity or Instinct."[60] What one trusts when one trusts oneself deeply is that inner quality of potentiality which, a generation after Emerson, nineteenth-century privacy advocates thought that privacy protects and, as I have argued, produces. This potentiality requires both personal and social resources of oblivion (what Emerson calls "that deep force," "without calculable elements," "the last fact behind which analysis cannot go") in order to have a reality that we can encounter and with which we can become acquainted without deluding ourselves or translating it into the form of unambiguous knowledge or information that extinguishes it.[61]

The form of self-trust grounded in oblivion helps us to deal with the pretty radical vulnerability of having a self at all. When it comes to living my own life, my powers of autonomy, self-knowledge, and the rest are limited. I come up short and have my blind spots. I let myself and others down in myriad ways. I am the mercy of myself. To go on living as a self-directed agent requires not being self-deluded about such failures, but also something like a background belief that such failures are neither definitive nor exhaustive of my potential in those domains and others. The connection here to the moral idea behind the right to be forgotten is evident. In much the same way that K's confidence in the possibility of forming new acquaintances oblivious of his past gave him the sense that his future was open and therefore that his life was worth the trouble it takes to live it, confidence in my own internal oblivion gives me a lived and living sense of inner potentiality that is inexhaustible by the facts of what my life has been like so far. (We can easily picture K turning his imagination to distant lands with no Internet, or to the inexhaustibility of his own inner life, just as Funes turned his head to the unknown quarter of dark houses.) This marks one important difference between self-trust and the trust of others: if someone fails us time and again, we might come to distrust that person on the whole—that is, not just with regard to the domains in which they failed to meet our expectations—and although that particular relationship might suffer, our lives might in the end go better. But if I fail myself repeatedly, I cannot stop trusting myself in the same way without suffering a serious debilitation of agency and self-esteem. We need the belief that we can be different going forward, all evidence to the contrary notwithstanding. Such belief is not a necessary psychological feature of human beings, but is built upon certain social and ideological foundations, among which are the various means for producing individual and social oblivion we have encountered so far. The idea that there are areas of ourselves to which neither we nor anyone else have access is essential for sustaining this belief across a lifetime.

The connection of oblivion and trust scales up from individual to common goods much like the depth-giving "darker ground" did. There can be no trust of any sort without some limits to knowledge; this is just what distinguishes trust from certainty or reliance. Therefore, we ought to think that the form, extent, and quality that barriers to what can be known about people take in a certain society would influence the amount and quality of trust available to the people living in it. Contemporary life abounds with examples in the form of tracking technologies. It is common today to use the GPS capabilities of smartphones to follow the movements of a child, friend, or lover through a city. Employers track their employees' keystrokes and Internet usage. Teachers monitor their students' eye movements. Parents keep track of their babies' heartbeats, oxygen levels, and more while they sleep. Each of these relationships reveals and perpetuates a lack of trust between the parties or between the individuals and the world at large (most notably in the case of the parents). Surveillance creates suspicion and therefore the urge to surveil. The seeming self-sustenance of this vicious cycle is another reason to resist the ideology of information that conflates privacy with secrecy. It may not have occurred to you that workers in an office were being suboptimally productive before you learned that they were being monitored. You don't have to be their manager or even work in the office for the idea to enter your head that because they are being monitored, maybe they have something to hide. Learning that other parents monitor their babies' vital signs at night makes me wonder whether I should, too. I wonder what I am missing, what hidden dangers lurk beneath the seemingly placid surface of my sleeping child.

The subjects of such tracking and monitoring are also deprived of the opportunity to be trusted. To the extent that the experience of being trusted is an important experience of personal development and moral self-worth—and I think that it is—the child who is tracked by her parents from her earliest opportunities for independence, whether in the physical world or online, and the worker whose every movement is monitored will be worse off because they are deprived of opportunities to be trusted. If this seems like small potatoes to those living in a world saturated by opportunities to track and be tracked, consider its effects in the aggregate. It is practically an axiom of human life that a society cannot function without trust, or at least that a free society cannot (Hobbes offers a different solution); therefore, societies are better or worse off depending on whether there is more or less "social trust."[62] Although social trust may have some irreducible gestalt or cultural aspects, by and large it will be the sum of all particular trust and distrust in a society: lovers abroad, workers on the job, students at their desks, parents in their worry. Local diminishments therefore have consequences for the common good. However, in our case the connection is much tighter, for any local diminutions of trust occasioned by tracking technology respond to the incentives not of individual circumstances but a broader technological change in the social practices of knowledge.

Privacy, forgetting, and other forms of oblivion provide the conditions for trust to grow in a society by generating opportunities to trust and be trusted as an aspect of everyday life. One could not hope to put it better than Joseph Kupfer: "Privacy is a trusting way others treat us, resulting in a conception of ourselves as worth being trusted."[63] Being trusted by others gives us a sense of self-worth and self-respect because it is connotes both a personal and generalized social judgment that we are deserving of trust. This sense is especially durable and valuable because it is, in Arendt's terms, objective: it reflects and expresses a judgment on behalf of the world at large that we are *worthy* of being trusted. The everyday experience of privacy is especially effective at conveying this sense of worth because the implicit judgment of trustworthiness is not restricted to a specific domain. When the world of others permits us to disappear into the oblivion of our private lives, we are not aware of being trusted to perform or abjure a specific set of actions (not beat our kids, yes feed them) but rather of being trusted more generally. In its holism, this experience expresses something greater than the sum of the individual reasons and instances of trust. It conveys a message along the lines of: they trust me to live my life because they think that I am worthy of doing so in my own way. Kupfer connects this experience to agency in a way that recalls both our arguments about K's confidence that his life was up to him, as well as the idea that epistemic unaccountability is vital for agency: "Armed with the sense of self-worth that turns upon belief in one's own trustworthiness, an individual has a kind of moral confidence in his choice-making abilities."[64] The loss of this sense of general trustworthiness explains why being surveilled, even in public and even, as in Michael Haneke's film *Caché*, by a single private citizen, can erode one's sense of belonging to or being at home in the collective social body.[65]

Much of what can be said about the good of social opportunities to be trusted also applies to the availability of opportunities to trust. Having the experience of trusting others is certainly part of the process by which children develop into morally mature adults and citizens, whatever else that entails.[66] And respecting others' privacy requires that one trust them, perhaps with regard to specific actions, but more commonly in the powerfully general way in which a person who is permitted to retreat into the oblivion of privacy is trusted by society. The varied domains of privacy present a multitude of quotidian, almost invisible opportunities to trust one's fellow citizens, which in turns helps one see them as morally independent equals. And since vulnerability is an ineradicable element of trust, privacy as a practice of trust also presents a practical reminder that we are always mutually vulnerable to one another as much as to ourselves. To be sure, history makes clear that respect for privacy is woefully insufficient on its own to generate a politics and ethics of mutual trust, respect, and solidarity in the face of such vulnerability. Even so, the civic, agential, and moral values inculcated by trusting against a background of oblivion contribute to a view of how the respect for privacy fits into an egalitarian, pluralistic politics.

We have seen by now that much of what is valuable about privacy depends upon the social reproduction of oblivion rather than other forms of opacity like secrecy or hiding. So, too, with trust. For privacy to convey the broad range of benefits that come from trust, it must refer to a zone about which there is not thought to be some hidden fact of the matter—a realm of oblivion, in other words, not secrets. The Russian proverb "trust, but verify" is rightly famous for its irony. To the extent that you verify, you do not trust, and to the extent that you are monitored, watched, or documented, you are not trusted. From this perspective we can redescribe the perennial privacy concern over the increasing documentation of individual lives as also stemming from the diminishment of the social bases of trust and trustworthiness: in a "record prison" one feels just as trusted as one would be in a real prison. Indeed, the elimination of opportunities for being trusted is the effective mechanism of Bentham's Panopticon. By contrast with the benefits to agency, self-conception, and citizenship that come from being allowed to disappear into private oblivion, from one's past self being allowed to disappear into mnemonic oblivion, and from allowing others to do the same, "monitoring behavior or collecting data on us projects a disvaluing of the self in question."[67] In this, the "record prison" coincides once more with the actual prison, where—as in Foucault's description of the panopticon and sexuality's imperatives of knowledge, Weber's analysis of the protestant ethic, and Goffman's sociology of the total institution—the absence of privacy is intended to erode an individual's sense of self-trust and trustworthiness grounded in realms of oblivion. The result, in Goffman's words that would reappear in Harcourt's critique of overwhelming publicity, is "the mortification of the self."[68]

Let us finally throw one last glance over our shoulder at the accusation that privacy is for those who have something to hide. The important point here is not that this slogan is wrong: this should be clear by now. The idea that we all have "skeletons in our mental closets," as Thomas Nagel puts it, and therefore that everybody has something to hide, degrades the quality of our trusting others by displacing the oblivion of privacy with the hidden information of secrecy.[69] It also corrupts the sense of our own trustworthiness by sending the message that one's privacy is respected not because one is fundamentally trustworthy, but because of the negative collective consequences that would ensue if everyone's shameful secrets were available for public view. Most importantly, by assuming that what is private is whatever "we all have to hide," privacy loses its capacity to give depth and strengthen trust in oneself and others. If we understand others' privacy merely as a shield for keeping information from us, we will be inclined to feel slighted, worried, or curious, rather than respectful and trusting. Even slight or incremental losses in this respect are serious—and grave in their accumulation—for they mean that the world of human activity is therefore less open, free, and humane, and that we have become shallower, deprived of the depth that makes our lives interesting, meaningful, and worth the often great trouble that it takes to live them.

POSTSCRIPT

When I give a talk on privacy, or when I discuss the ethical dilemmas of the digital age with friends and family, I almost always get the question, "So what should we do?" We are hungry for solutions to the challenges of new technologies. Yet many of us feel helpless to address these challenges on our own, perhaps because they concern massive and fast-moving political and economic forces, but also because they often involve rapidly changing technologies that we have trouble keeping up with or even understanding. People are right to want answers. It is here that I run the risk of disappointing readers and aunts alike. For in my experience, when someone asks what is to be done, they are looking for technical solutions to ethical and political problems. By "technical" I do not just mean new technologies, but also what laws and regulations we should pass, what adjustments to our "privacy settings" online we should make, what we can do to avoid being tracked in public, and so on. These approaches to the problem are important, to be sure, although these days, technical responses are out of date almost as soon as they are suggested, coded, printed, or signed into law.

This book has offered a different kind of answer to the crisis of privacy in the digital age, in recognition that even the cleverest technical solutions, be they novel devices or sweeping new regulations, are limited by the conceptual frameworks available to them. These frameworks are what render a particular set of problems visible in the moral and political landscape— they are what make problems problematic, in other words. Each framework also suggests a certain range of responses to the problems it reveals. Yet because our conceptual frameworks draw our attention to certain aspects of an irreducibly complex political and ethical situation, they inevitably obscure others. Even the most creative problem solver cannot think beyond the limits of the frame she uses to conceptualize the issue, for to do so would be to think about a different problem entirely. The issue is especially serious today, when the dominant attempts to understand, respond to, and defend against the diminution of privacy in the information age adopt the conceptual framework of privacy's opponents. The adoption is subtle, and far from total, but it links privacy's defenders and deprivers on at least one highly significant point of common ground. This is, as we have seen, the focus on information and its naturalization in human affairs as somehow prior to the question of privacy—a significant mistake in the recent history of thought about privacy, and one which this book hopes to remedy.

So, this book does suggest solutions to the challenges of the digital era, after all, not in the realms of law and technology, but in those of value and understanding. It offers a new way to approach the challenges of our time, which may then guide political and technical action. I have also sought to develop a view of human well-being in which oblivion and barriers to information play a vital role. This led to an argument for an ethics of self-detachment, and for a politics that permits, enables, and encourages this experience and distributes it equitably among all citizens. As a society, we have become so caught up with technical solutions to technological problems that we ignore the crucial ethical and political questions at hand. We have, as I suggested in the introduction, lost sight of the deeper and more important questions that we should be asking about the structure of human life and the role that technology plays in it. I have aimed to invite questions of value where perhaps there were none before—or where they have long since fallen silent—to challenge us to think differently about the importance of privacy in an age hyper-fixated on the power and profitability of information. Most of all, I have tried to show us and our world to ourselves anew, in all our messiness and strangeness, so that we might see with new eyes the vital role that oblivion plays in human affairs, its depth-giving, uncanny refulgence.

NOTES

ACKNOWLEDGMENTS

INDEX

Notes

Introduction

1. "Data subject" is how the European Union's omnibus General Data Protection Regulation of 2016 refers to "an identified or identifiable natural person" (Art 4 [1]). This description of the person that privacy protects expresses a conception of what that person consists in, and its migration into the literature on privacy as a description of what privacy protects is but one example, of many cited in this book, of the dangers to our understanding of the person that come from valuing privacy in the wrong way.

2. Deborah Lupton, "You Are Your Data: Self-Tracking Practices and Concepts of Data," in *Lifelogging: Digital Self-Tracking and Lifelogging—between Disruptive Technology and Cultural Transformation*, ed. Stefan Selke (Wiesbaden, Germany: Springer Fachmedien Wiesbaden, 2016), 61–79; *The Quantified Self: A Sociology of Self-Tracking* (Cambridge: Polity, 2016).

3. Werner Herzog and Chris Heath, "Mad German Auteur, Now in 3-D!," *GQ*, April 29, 2011, https://www.gq.com/story/werner-herzog-profile-cave-of-forgotten-dreams.

4. Bernard E. Harcourt, *Exposed: Desire and Disobedience in the Digital Age* (Cambridge, MA: Harvard University Press, 2015), 175–176.

5. Harcourt, *Exposed*, 176.

6. Shoshana Zuboff argues that the political economy of *surveillance capitalism*, a term she brought to popular consciousness, is not primarily concerned with harvesting data about us but rather producing it where there had been none before. If surveillance capitalism really does refer to a novel economic formation, as Zuboff argues, then we would expect it to be accompanied by new ideological constructs. The ideology of information—that is, the idea that much or all of human life may be unproblematically understood in terms of information, and that privacy safeguards information rather than protecting against its creation—is exactly what one might expect to emerge from the political-economic forces of information extraction that Zuboff describes in *The Age of Surveillance Capitalism: The Fight for a Human Future at the New Frontier of Power* (New York: Public Affairs, 2018). Indeed, the very notion that data is something that companies can "harvest," as if it had an ontologically

independent existence prior to the actions of those corporations—who manufacture data, not reap it—is itself a product of the ideology of information.

7. The waters of understanding have been muddied in this case by an overreliance on the views of lawyers, judges, and legal academics on the value of privacy—that is, as a social phenomenon or human interest, beyond the narrower question of the meaning of privacy in the law. This is not to impugn the work of many scrupulous scholars, but rather to point out that the development of ideas in the law is governed by disciplinary constraints concerned with things other than tracking the truth about privacy. It is a common feature of work on privacy by all sorts of scholars, and not just lawyers, that it tends to take what judges say on the matter pretty seriously and to assume that an adequate conceptual account of privacy must account for both its ordinary usage and the eccentric, somewhat parochial usage particular to US constitutional law. However, unlike philosophy or popular moral discourse, jurisprudence is restricted by internal principles of stare decisis, adherence to precedent, and fidelity to written or customary law. Judges who wish to describe the right to privacy and what it protects are significantly constrained by what previous judges have said, as they were in turn, for reasons particular to their profession; lawyers and scholars engaging in *that* discussion are likewise constrained. This nearly creates a problem if the starting place for that chain of reasoning is itself already divorced from the ordinary use of privacy. It only *nearly* creates a problem because it is not an issue for judges whether their language tracks the ethical and political realities of the society under their jurisdiction, only that their language tracks some plausible interpretation of the law. The situation becomes problematic when writers from other disciplines take the history of privacy jurisprudence as tracking in any reliable way the answer to the question of why privacy matters in an ethical, political, or sociological sense. (Frequent equivocation between moral and legal rights to privacy is another source for this confusion in the literature.) That said, I will look to several legal documents in my account of the history of privacy, but only those which argued for the recognition of a new legal right to privacy in response to a perceived moral demand. These documents are useful (and authoritative) in just the same way that newspaper columns or books on the subject are: they reveal what people thought at the time about the moral value of privacy and the threats that it faced.

8. Jeroen van den Hoven, Martijn Blaauw, Wolter Pieters, and Martijn Warnier, "Privacy and Information Technology," *The Stanford Encyclopedia of Philosophy* (Summer 2020), ed. Edward N. Zalta, https://plato.stanford.edu/archives/sum2020/entries/it-privacy/. The encyclopedia's other article on privacy says the same thing at much greater length and without the concision of the entry from which I quote in the body of the text. Beate Roessler and Judith DeCew, "Privacy," *The Stanford Encyclopedia of Philosophy* (Winter 2023), ed. Edward N. Zalta and Uri Nodelman, https://plato.stanford.edu/archives/win2023/entries/privacy/.

9. The phrase "contextual integrity" is Helen Nissenbaum's, who also views the right to privacy as a "right to [the] appropriate flow of information." Helen Fay Nissenbaum, *Privacy in Context: Technology, Policy, and the Integrity of Social Life* (Stanford, CA: Stanford Law Books, 2010), 145. See also Herman Tavani, "Informational Privacy: Concepts, Theories, and Controversies," in *The Handbook of Information and Computer Ethics* (Hoboken, NJ: Wiley,

2009) 131–164; Andrei Marmor, "What Is the Right to Privacy?," *Philosophy and Public Affairs* 43, no. 1 (2015): 3–26; and Julie E. Cohen, "What Privacy Is For," *Harvard Law Review* 126, no. 7 (2013): 1904–1933.

10. Van den Hoven, et al., "Privacy and Information Technology" (citing Jeroen van den Hoven, "Information Technology, Privacy, and the Protection of Personal Data," 2008). This substantial narrowing of privacy's concept and moral force to mere "data protection" is a consequence of its prior reformulation as "informational privacy."

11. E.g., Nissenbaum, *Privacy in Context*; Marmor, "What Is the Right to Privacy?"

12. Again, I am convinced that we need strong rights and regulations of "data protection" to protect against the many ways that information about us can be misused or used to harm us. Many of these are enshrined in laws like the EU's general data protection regulation (GDPR), and I think they are necessary and important. I am not arguing against rights concerned with data or information, only that conflating the importance of privacy with that of data obscures values that stand in opposition to data and information.

13. By "knowledge" here I mean the type of knowledge associated with information: what philosophers call "propositional knowledge" or "knowledge-that." We will have more to say about this in the chapters to come.

14. Joshua Rothman, "Virginia Woolf's Idea of Privacy," *New Yorker*, July 9, 2014 (available at https://www.newyorker.com/books/joshua-rothman/virginia-woolfs-idea-of-privacy).

15. "'Is There Any Privacy?,'" *Hartford Daily Courant*, October 3, 1874; "The Decay of Privacy," *Boston Daily Globe*, January 19, 1922; "'Is the End of Privacy Coming to Human Kind?,'" *Boston Daily Globe*, September 30, 1928. The history of this lament is discussed in chapters 1 and 3.

16. For various reasons, scholars also tend to periodize the information age with the development of computers or the Internet and the opening of cyberspace. See, e.g., Yoneji Masuda, "Image of the Future Information Society" in Frank Webster et al., eds., *The Information Society Reader* (London: Routledge, 2004), 15–20; and Ester Dyson, George Gilder, George Keyworth, and Alvin Toffler, "Cyberspace and the American Dream," in *The Information Society Reader*, 31–42. I do not wish to enter into these debates, which often focus on the primary economic organization of a society, and, as David Edgerton, Theodore Roszack, Kevin Robins, Frank Webster, and other historians of science have argued, can be both wrong and misleading. All historical periodizations are, at best, rough heuristics; at worst, they are ideological and misleading. So I offer my rough and ready view of "the information age" as a way of marking out the huge increase in the documentation of ordinary life stretching from the second half of the nineteenth century to the twenty first in order to show that the advocacy for privacy rights and interests, as a reaction to that increase, is nearly coextensive with the period. For a good overview of the periodizations of the "information age" and their critiques, see generally *The Information Society Reader* (London: Routledge, 2004).

17. David Edgerton's *The Shock of the Old* (Oxford: Oxford University Press, 2006) offers a much-needed tonic to the constant assertions of the revolutionary and novel aspects of our current moment in the history of technology.

18. Ralph Ellison, *Invisible Man*, 2nd ed. (New York: Vintage, 1995), 3. Ellison's characterization of the white world's blindness as a result of the perceptual habits inculcated in a

racist society—"a matter of the construction of their *inner* eyes, those eyes with which they look through their physical eyes upon reality"—anticipates the arguments in chapter 1 about the photographic revision of sight. See also Charles Mills's "White Ignorance" from *Black Rights/White Wrongs* (Oxford: Oxford University Press, 2017), which may also be fruitfully accessed alongside several other approaches to the subject in Shannon Sullivan and Nancy Tuana, *Race and Epistemologies of Ignorance* (Albany: State University of New York Press, 2007).

19. Here a proponent of the "informational privacy" views might say: But what's being protected in these cases *is* information—it's the information about the location of the fishing hole, or the location (or appearance of one's body). To which I say: That's precisely why they're secret fishing holes or hideaways, and not private ones. On secrets, see Sissela Bok, *Secrets: On the Ethics of Concealment and Revelation* (New York: Pantheon, 1983). Bok is, as far as I can tell, the only one to analyze the relation between secrets and privacy and to argue for the importance of not conflating the two (in her case, for understanding secrecy). However, we differ on what *privacy* means, specifically with regard to its relation to information and incompatibility with hiding. Indeed, it is extremely curious that while Bok takes pains not to elide secrecy and privacy, she does not insist on the distinction with hiding. For instance, "Privacy and secrecy overlap whenever the efforts at such control [i.e., over "what one takes—however grandiosely—to be one's personal domain"] rely on hiding" (11). Chapter 3 will argue at length for the distinction between privacy and hiding, and the great importance of not forgetting it; the distinction between secrecy and privacy is one for which I argue throughout the book.

20. Shakespeare, *Richard III*, Act 3, Scene 7, lines 130–131.

21. See also the Treaty of Edinburg (1560): "All things done here against the laws shall be discharged, and a law of oblivion shall be established." Both Acts appear in Lewis Hyde, *A Primer for Forgetting: Getting Past the Past* (New York: Picador, 2020), 69.

22. Hyde, *Primer for Forgetting*, 69.

23. See, e.g., Jake Pearson, "NYers Furious over Photos Taken through Windows," *Associated Press*, May 17, 2013, https://news.yahoo.com/nyers-furious-over-photos-taken-132312 962.html; J. Renwick, *Foster v. Svenson*, No. 2015 NY Slip Op 03068 [128 AD3d 150] (New York Appellate Division, First Department July 1, 2015); "France: Intrusive Olympics Surveillance Technologies Could Usher in a Dystopian Future," Amnesty International, March 20, 2023, https://www.amnesty.org/en/latest/news/2023/03/france-intrusive-olympics-surveillance -technologies-could-usher-in-a-dystopian-future/; Pete Grieve, "To Save Money on Insurance, Drivers Are Agreeing to 'Incredibly Intrusive' Monitoring Technology," *Money*, https:// money.com/usage-based-car-insurance-gaining-customers/ (accessed July 30, 2023).

24. Gay Talese, *The Voyeur's Motel* (New York: Grove Press, 2016), 26.

25. E.g., Thomas Nagel, "Concealment and Exposure," in *Concealment and Exposure: And Other Essays* (Oxford: Oxford University Press, 2002), 3–26.

26. Confidentiality concerns the communication of secrets. The description of confidential information as "privileged" refers to the fact of a certain arrangement for knowledge of the secret that includes some but excludes others. So obviously, confidentiality requires at least two parties. This is also because confidentiality describes a relationship of trust (or at

least reliance), in some cases based in the good of one party (for instance, with medical information), in others justified by utilitarian considerations about maximizing systemic benefits and minimizing harm (for instance, lawyer-client confidentiality). Although confidentiality is typically understood today in quasi-contractual terms, it has its basis in the idea that confidentiality gives one the (affective) confidence that one's information will not be shared. We need confidentiality laws or professional codes of ethics because we don't know these people well enough to trust them. One shares a secret *in confidence*; one shares one's medical information with a new provider in the confidence that they will not disclose it to anyone without one's permission. This is the connection between confidentiality and the older use of the word *confidence* to indicate something like trusting assurance: you *confide in* someone (a *confidant*) who inspires trust that your secret will be safe with them.

27. Willis Barnstone and Jorge Luis Borges, "Thirteen Questions: A Dialogue with Jorge Luis Borges," *Chicago Review* (Winter 1980): 16.

1. Photography and the Invention of Privacy

1. Svenson gallery materials, quoted in J. Renwick, *Foster v. Svenson*, no. 2015 NY Slip Op 03068 [128 AD3d 150] (New York Appellate Division, First Department July 1, 2015).

2. Jake Pearson, "NYers Furious over Photos Taken through Windows," *Associated Press*, May 17, 2013, https://news.yahoo.com/nyers-furious-over-photos-taken-132312962.html.

3. *Foster v. Svenson* (2015).

4. Take a moment to look around while holding the thought in your mind that everything you see is the result of light bounding off the world and streaming into your eyes. Now try to maintain that awareness for a single day. A minute or two is enough to give me vertigo, so uncanny is the world that pours into my sight when compared to the world into which my gaze goes out.

5. Warren and Brandeis, "The Right to Privacy," *Harvard Law Review* 4, no. 5 (Dec. 15, 1890), 195.

6. Godkin, "The Rights of the Citizen: IV—To His Own Reputation," *The Nation*, December 25, 1890, 496–497; Warren and Brandeis, "The Right to Privacy," 193.

7. See the *Oxford English Dictionary* entry on "privy," a real treasure trove.

8. *J. Campbell, Rep. Cases Nisi Prius vol. 3 81* (quoted in the *O.E.D.* entry on "privacy).

9. Nathaniel Hawthorne, *Collected Novels: Fanshawe, The Scarlet Letter, The House of the Seven Gables, The Blithedale Romance, The Marble Faun* (New York: The Library of America, 1983), 212.

10. For instance, this 1740 notice from the Pennsylvania Gazette in which privacy is a synonym for confidentiality: "For the satisfaction of the Publick, shortly will be printed, a certain private Letter sent under Seal to William Cosby, Sheriff of New-York, which by a certain Grand Jury upon their solemn Oath, was Declared to be a false, malicious, scandalous, defamatory, libelous Writing, published and caused to be published (notwithstanding its Privacy) together with the Reasons upon Oath for Writing the fame." *The Pennsylvania Gazette*, December 4, 1740. It is precisely to distinguish the modern value of privacy from such previous usages that early privacy advocates continually insisted on its novelty.

11. "The Sacred Privacy of Home," *The Independent . . . Devoted to the Consideration of Politics, Social and Economic Tendencies, History, Literature, and the Arts (1848–1921)*, August 8, 1850.

12. Accounts of this development can be found, for instance, in Jerrold E. Seigel, *The Idea of the Self: Thought and Experience in Western Europe since the Seventeenth Century* (New York: Cambridge University Press, 2005); Richard Sennett, *The Fall of Public Man* (New York: Norton, 1996); Charles Taylor, *Sources of the Self: The Making of the Modern Identity* (Cambridge, MA: Harvard University Press, 1989). About urbanization: In the nineteenth century, the population of London grew from 860,000 to 5 million; New York City exploded from 60,000 to 3.4 million. (Sennett, *Fall of Public Man*, 50–51; Ira Rosenwaike, *Population History of New York City* (Syracuse, NY: Syracuse University Press, 1972); The City of New York, Population Figures (available at https://www1.nyc.gov/assets/planning/download /pdf/data-maps/nyc-population/historical-population/1790-2000_nyc_total_foreign_birth .pdf). Life in these new urban centers was conducted in a thick milieu of strangers living in close proximity, constantly mixing, yet without the pre- and early-modern contexts in which a stranger could be placed at a glance (i.e., one could be easily identified as either from this village or foreign to it) (Sennett, *Fall of Public Man*). Many found themselves sharing space with or looking into the windows and overhearing the conversations of others about whom they knew nothing other than what could be gathered from the immediate encounter. It was against this background that the modern demand for privacy emerged. For example, this 1874 column on "The Right of Privacy" laments the end of a brief period in which the new, chaotic cities furnished the first conditions for privacy to flourish:

> There was a time when the dwellers in rural hamlets were in one respect miserable above all others. A system of mutual explorage [sic] practically abolishes privacy in such neighborhoods now as then. A villager can do nothing that can by any ordinary means be kept from the public. He can say nothing, even within the walls of his own house, that may not in less than six hours after its utterance be discussed in the barrooms and corner groceries. His strictly private affairs are regarded as public property. . . . This sort of thing once pertained to country towns, and time was when if a man desired to become a recluse he had his choice, whether to pitch his tent in a great city or in the wilderness. In either place he was certain of that quiet which favors calm contemplation and the length of years. The privilege of being a good citizen without ostentation, and without being drafted into common conversation on slight occasions, and on no occasions, was then possible. It is no longer possible amid the haunts of man."
> ("The Right of Privacy," *North American and United States Gazette*, July 11, 1874)

13. See, e.g., Taylor, *Sources of the Self: The Making of the Modern Identity*; Seigel, *The Idea of the Self*; Nancy L. Rosenblum, *Another Liberalism: Romanticism and the Reconstruction of Liberal Thought* (Cambridge, MA: Harvard University Press, 1987); Sennett, *Fall of Public Man*; Harry G. Frankfurt, "Freedom of the Will and the Concept of a Person," in *The Importance of What We Care About: Philosophical Essays* (Cambridge: Cambridge University Press, 1988), 11–25.

14. See, e.g., Walter Benjamin, "Paris: Capital of the Nineteenth Century," in *Reflections: Essays, Aphorisms, Autobiographical Writing* (New York: Schocken, 1986).

15. For an extended account of these developments, see Sennett, *Fall of Public Man.*

16. "The Sacred Privacy of Home." The column begins with a litany of goods and freedoms associated with spontaneity and authenticity, which are protected by privacy not primarily against perception but against the constricting *internalized* gaze of social mores (which is, of course, perpetuated and enforced by actual perception): "One of the most attractive features of a good home is its privacy. There conversation is conducted with the freedom of mutual confidence and affection; there the meal is divested of all formality and constraint, and made truly social; there dress is unstudied as to fashion or its material." What privacy gives the occupants of this "sacred home" is the confidence that no matter what they do or say or how they dress, it will not be documented or pass beyond the bounds of the moment as it is lived. Privacy clearly is not the norm that prevents such transmission; that is achieved by "mutual confidence," trust, and the reliable opacity of walls. By contrast, privacy is depicted as a state of affairs characterized by the absence of information about individuals; this state of affairs may be produced by those barriers to perception but is not equivalent to them. This condition of privacy, in turn, is what gives individuals the confidence necessary to give themselves over to the unpredictable vagaries of spontaneous behavior—and, to anticipate a major line of argument in the rest of the book, provides a time and space for personality and self-conception to come apart. The persistence of this nineteenth-century view into the present day can be seen everywhere—note for instance, how closely Thomas Nagel echoes "The Sacred Privacy of Home": "The public gaze is inhibiting because, except for infants and psychopaths, it brings into effect expressive constraints and requirements of self-presentation that are strongly incompatible with the natural expression of strong or intimate feeling." Thomas Nagel, *Concealment and Exposure: And Other Essays* (Oxford: Oxford University Press, 2002), 18.

17. Ralph Waldo Emerson, "Self Reliance," in *The Essential Writings of Ralph Waldo Emerson*, the Modern Library Classics (New York: Modern Library, 2000), 141 (emphasis supplied).

18. This lineage is particularly evident in the development of the legal right to privacy in Germany. The German Constitutional Court bases the fundamental right to privacy in a "general right of personality" ("Allgemeines Persönlichkeitsrecht"), which is in turn founded on two fundamental rights enshrined in its 1949 constitution: the right to human dignity and the right to the free development or *unfolding* of personality—*die freie Entfaltung seiner Persönlichkeit.* The idea of a personality that can be unfolded—and whose unfolding corresponds to an aspect of human well-being deserving of rights protection—is distinctly Romantic and, to a significant extent, German in origin. For instance, here is Friedrich Schiller: "Personality [*Persönlichkeit*], considered in itself and independently of any sense-material, is merely the disposition for potentially infinite expression" (Schiller, *On the Aesthetic Education of Man*, [London: Penguin Classics, 2016], 40). I invoke Schiller because of the central role of his thought in the development of the Romantic ideology of personality that would become foundational for modern privacy—in contrast to, say, Kant's moral personality ("moralische Persönlichkeit")—but also because of his formative influence on Louis Brandeis, who read Shiller all his life until his sight failed, when it had it read to him (Melvin I. Urofsky, *Louis D. Brandeis: A Life* [New York: Pantheon, 2009], 15, 35). It is

almost impossible to read Brandeis's account of "inviolate personality" without hearing the influence of German Romantics and proto-Romantics, like Schiller. The influence of Schiller's idea of personality was, of course, everywhere in Warren and Brandeis's post-Emersonian Boston.

19. Charles Darwin, *The Expressions of the Emotions in Man and Animals* (New York: D. Appleton and Company, 1898).

20. Darwin, *Expressions of the Emotions*, 75.

21. Arthur Schopenhauer, "On Physiognomy," in *Religion: A Dialogue, and Other Essays* (S. Sonnenschein and Co., 1891).

22. "'Is There Any Privacy?,'" *Hartford Daily Courant*, October 3, 1874; "No More Privacy," *Kansas City Star*, January 28, 1889; "The Decay of Privacy," *Boston Daily Globe*, January 19, 1922; "'Is the End of Privacy Coming to Human Kind?,'" *Boston Daily Globe*, September 30, 1928.

23. John Gilmer Speed, "The Right of Privacy," *North American Review (1821–1940)* 163, no. 476 (1896): 64–74; George Corkhill, "Portrait Right," *Washington Law Reporter* 12 (1885 1884): 353; John Bascom, "Public Press and Personal Rights," *Education* 4, July 1884: 604–605.

24. Warren and Brandeis, "The Right to Privacy," 195. Although "The Right to Privacy" presents the case for a new legal right to privacy, the argument for the necessity of such a right takes place in the register of moral and political theory. It is in this sense that we will read it here. Although the article's invocation of law and precedent sought to provide pragmatic legal justification for the new right to privacy—that is, reason for thinking that such a right is already entailed by common law precedent, given that no legislatures had yet enacted one—reference to the common law will not give reason to think that such a right is necessary or desirable. We can dream up all sorts of rights that are logical extensions of the law as it is, but this will not give us reason to think that we have (or ought to have) such rights in the first place. Only moral and political argument can do that. So we find that the heart of "The Right to Privacy" is an argument for a certain view of human well-being, the interests it generates, and the threats facing those interests. The same goes for the court decision in *Pavesich* quoted below, or the opinion in Arne Svenson's case.

Of all the elements of moral life, rights seem most amenable to historical study. Notions of virtue, normativity, and basic goods permeate our lives more deeply yet can reach their full expression without necessarily leaving a trace in the historical record. But there is something about rights, beyond the fact that they emerged and developed in an age of increasing literacy and publication, that seems to demand that one's views be written down and printed. Published advocacy for novel rights is especially useful for anyone interested in tracing the history and future of ethical concepts. This is because normative discourse about new rights must make a display of the past and future of the very concept on offer. Just as one cannot reasonably argue for the acceptance of a new right without reference to some preexisting ethical culture and concepts by which it appears morally legible and warranted, neither would a new right be plausible if one could not give an account of how things would be better off with it in the world. Successful accounts of new rights tend to succeed in both these ways.

25. Warren and Brandeis would have been familiar, for instance, with the entry on eavesdropping in Blackstone's *Commentaries*, which would have been more than sufficient to protect against enterprising snoops without the added trouble of arguing for a brand-new right: "*Eaves-droppers*, or such as listen under walls or windows or the eaves of a house, to hearken after discourse, and thereupon to frame slanderous or mischievous tales, are a common nuisance and presentable at the court-leet, or are indictable at the sessions and punishable by fine and finding sureties for their good behavior." William Blackstone, *Commentaries on the Laws of England. In Four Books*, vol. 2 (J. B. Lippincott Company, 1893), 168. Blackstone is quoted approvingly in a US criminal law text published in Boston in 1872. Joel Prentiss Bishop, *Commentaries on the Criminal Law.*, 2 v. (Boston: Little, Brown and Company, 1872), 657–658.

26. See, e.g., Bill Jay, *Cyanide and Spirits: An Inside-Out View of Early Photography* (Munich, Germany: Nazraeli, 1991).

27. Warren and Brandeis, "The Right to Privacy," 196. In chapter 3 we will return to the role that the new mass media played in the development of modern privacy, so here let me simply state the following. We might think that the huge readership of emerging mass media raised to a new magnitude the danger of having true information spread around, but this wouldn't account for the broadly shared sense that there was something radically new and modern about the mode of invasion distinct from its consequences. Nor would it explain why *photography* was of special concern rather than just newspapers. It is much easier to paint someone in a bad (but nevertheless true) light with words than with a photograph— for one thing, you needn't catch them in the act. Anyway, "The Right to Privacy" was published seven years before the first half-tone photograph would appear in US newsprint. Pictures would not play a significant role in newspaper journalism until after the first World War. Michel Frizot, "The All Powerful Eye," in *A New History of Photography*, ed. Michel Frizot (Köln, Germany: Könemann, 1998), 365.

28. Edgar Allan Poe, "The Daguerreotype," in *Classic Essays on Photography* (New Haven, CT: Leete's Island Books, 1980), 38.

29. Oliver Wendell Holmes, "The Stereoscope and the Stereograph," in *Classic Essays on Photography* (New Haven, CT: Leete's Island Books, 1980), 82.

30. Edward Weston, "Seeing Photographically," *Complete Photographer* 9, no. 49 (1943): 3200–3306.

31. Holmes speaks of taking images from the world as a hunter takes skins from an animal. Holmes, "Stereoscope and the Stereograph," 81.

32. Poe, "Daguerreotype," 37–38.

33. The rapidity with which we adapted to "photographic distortion" and accepted it as natural and indexical to the three-dimensional world is indicative of the change in vision and subjectivity for which I argue in this section. The suddenness of this change also lends support to the view of just how destabilizing it was. Here is how William Ivins describes it:

> As people became habituated to absorbing their visual information from photographic pictures printed in printers' ink, it was not long before this kind of impersonal visual record had a most marked effect on what the community thought it saw with its own eyes. It began to see

photographically, it stopped talking about photographic distortion, and finally it adopted the photographic image as the norm of truthfulness in representation. A faith was put in the photograph that had never been and could not be put in the older handmade pictures. There have been many revolutions in thought and philosophy, in science and religion, but I believe that never in the history of men has there been a more complete revolution than that which has taken place since the middle of the nineteenth century in seeing and visual recording. (William Mills Ivins, *Prints and Visual Communication* [Cambridge, MA: MIT Press, 1969], 94)

34. *The Pencil of Nature* was the title of the first commercially published book illustrated with photographs, first printed in 1844. The book's author, William Henry Fox Talbot, was one of the prime innovators and early theorists of the photograph; he invented the "calotype" emulsion, one of the earliest positive-negative photo processes and a precursor to those that would make complaints like Warren and Brandeis's possible. In their catalogue, the Metropolitan Museum of Art in New York calls Fox Talbot's book "a milestone in the art of the book greater than any since Gutenberg's invention of moveable type," echoing the epochal shift in the epistemology of everyday life that came with the photograph. In a notice about the production of the fourth volume of *The Pencil of Nature*, *The Atheneum* referred to the project as "modern necromancy," expressing the contemporary sense that photographs could make even the dead speak and be present, which, as we shall see below, was central to the idea of the new photographic invasion ("New Publications"). "The mirror with a memory" comes from Oliver Wendel Holmes's article "The Stereoscope and the Stereograph," published in 1859.

35. "The Right of Privacy," *North American and United States Gazette*, July 11, 1874.

36. Paul Valéry, "The Centenary of Photography," in *Classic Essays on Photography* (New Haven, CT: Leete's Island Books, 1980), 193–194.

37. Valéry, "Centenary of Photography," 196.

38. We might look forward to the discussion in chapter 4 by noting the similarity of this development to the rise of the Internet as a global archive and how it affected the ways that we come to know, and trust, others; even more so to chapter 5's discussion of the relationship of privacy and trust.

39. Warren and Brandeis, "Right to Privacy," 198 (emphasis supplied).

40. "Alarming Possibilities," *Atchison Daily Globe*, March 25, 1866.

41. "No More Privacy," *Kansas City Star*, January 28, 1889.

42. *Marian Manola v. Stevens & Myers*, New York Supreme Court. Warren and Brandeis were evidently following this case closely, as they cite not only the court case but three days of *New York Times* coverage. "'New York Times' of June 15, 18, 21, 1890." For coverage of the event and case, see "Will Not Be Photographed in Tights," *Chicago Daily Tribune*, June 13, 1890; "The Rights and Tights of an Actress," *Baltimore Sun*, June 19, 1890. See also Jessica Lake, *The Face That Launched a Thousand Lawsuits: The American Women Who Forged a Right to Privacy* (New Haven, CT: Yale University Press, 2016).

43. Warren and Brandeis, "Right to Privacy," 195.

44. *Pavesich v. New England Life Ins. Co.*—122 Ga. 190, 50 S.E. 68 (1905), 220.

45. Mark Poster, *The Mode of Information: Poststructuralism and Social Context* (Cambridge: Polity, 1990), 97–98 (emphasis supplied).

46. Kevin D. Haggerty and Richard V. Ericson, "The Surveillant Assemblage," *British Journal of Sociology* 51, no. 4 (2000): 614.

47. Anna Hedenus and Christel Backman, "Explaining the Data Double: Confessions and Self-Examinations in Job Recruitments," *Surveillance and Society* 15, no. 5 (2017): 640–654.

48. Jay, *Cyanide and Spirits*, 227.

49. Jay, *Cyandie and Spirits*, 227. See also Robert Mensel, "'Kodakers Lying in Wait': Amateur Photography and the Right of Privacy in New York, 1885–1915," *American Quarterly* 43, no. 1 (1991): 24.

50. "The Concealed Camera for Newspaper Men," *Photographic Times and American Photographer* 18, no. 70 (February 10, 1888).

51. Schopenhauer, "On Physiognomy."

52. "The Legal Relations of Photographs," *American Law Register*, 1829.

53. Warren and Brandeis, "Right to Privacy," 206 (emphasis supplied). This also explains why privacy developed when it did, rather than in the 1840s and 1850s when the daguerreotype and similar photographic technologies were commonly available in the US and Europe. Although photography was widespread in the industrialized West by midcentury, the first cameras were so large, and their plates so finicky and slow, that a photographer needed her subject to sit still for a long time in full light for the subject to appear on film. This is why everyone who appears in a photograph taken before the 1870s wears more or less the same expression on their faces: one had to hold perfectly still for 15–20 seconds or their features would come out blurry in the print. Surreptitious and candid photography was impossible until a series of technical advances in the last quarter of the century made possible the sort of camera with which we are familiar today: small, portable, workable by amateurs, and capable of the short exposure times that permitted an operator to capture action and candid expressions.

54. Schiller, *On the Aesthetic Education of Man*, 40. On Brandeis's lifelong engagement with Schiller, see Urofsky, *Louis D. Brandeis*, 15, 35.

55. Emerson, "Self-Reliance." And William James: "The Mind is at every stage a theatre of simultaneous possibilities." William James, *The Principles of Psychology*, vol. 1 (New York: Henry Holt, 1950), 288. Of course, the value of such potentiality was not absolute. For human beings to become individuals, they must make something out of this potential, to appear to others and to themselves as being some way rather than another.

56. In other words, it was a good thing to "contain multitudes," as Walt Whitman put it, not just because it meant one could develop an individual personality out of the polyphony, but also because containing multitudes is a valuable human trait as such.

57. Warren and Brandeis, "Right to Privacy," 199.

58. Rosenblum, *Another Liberalism*, 50.

59. Of course, it is important that individuals be able to exercise some control over the boundaries of their privacy, and this exercise will to some extent serve to *maintain* the boundary between privacy and publicity. However, what makes privacy *privacy* (as opposed to publicity but also secrecy, confidentiality, hiding, etc.) will not be a feature of such control. Nor will the value of privacy derive from the importance of exercising some control in maintaining its epistemic barriers; rather, it's the other way round.

60. Warren and Brandeis, "Right to Privacy," 195–198.

61. In the Emersonian milieu of nineteenth-century Boston, it was common to refer to the individual and her appurtenances as sacred. Compare Warren and Brandeis's defense of privacy with the words of Amos Bronson Alcott, printed in an 1841 edition of *The Dial:* "Individuals are sacred. The world, the state, the church, the school, all are felons whensoever they violate the sanctity of the private heart" (Amos Bronson Alcott, "Orphic Sayings," *The Dial* 1, no. 1 [July 1840]: 85–98). The nineteenth century would see the language of the sacred applied with increasing frequency to sites associated with self-knowledge and knowledge of others: personality, of course, and the home, which even today is still referred to as a *sanctum*, but also the telegraph network, the mail, newspapers, and anonymized census data. The idiom of sacredness was so commonly applied to the individual and its environs that by 1911 William James would lament that such "phrases are so familiar that they sound now rather dead in our ears" (William James, *Talks to Teachers on Psychology, and to Students on Some of Life's Ideals* (New York: Norton, 1958).

62. Warren and Brandeis, "Right to Privacy," 201.

63. For an account of how we might understand such a change, see the discussion of Hannah Arendt's social ontology of privacy in chapter 5.

64. Warren and Brandeis, "Right to Privacy," 205.

65. Warren and Brandeis, "Right to Privacy," 201.

66. Warren and Brandeis, "Right to Privacy," 201.

67. For example, police have used real-time brain scanning to see whether suspects' neural activity displayed a "recognition pattern" when they were shown a picture of the weapon used in a recent murder. Nita A. Farahany, *The Battle for Your Brain: Defending the Right to Think Freely in the Age of Neurotechnology*, 1st ed. (New York: St. Martin's, 2023), 4, 78–79.

68. Farhany, *Battle for Your Brain*, 80. Note that scanning one's brain "to see what one really thinks" is to produce what one really thinks, in a medium thought to be more reliable than verbal testimony, and not to access or read it.

69. Sue Halpern, "The Bull's-Eye on Your Thoughts," *New York Review of Books*, November 2, 2023.

70. Farahany, *Battle for Your Brain*, 84.

71. This theme runs throughout Farahany's book, which means that it goes without saying that she does not problematize the existence or creation of information *as such.* Even her discussions of the proper limits of self-knowledge presuppose the existence of *facts* or *information* about oneself that perhaps one would be better off not knowing. This sort of self-ignorance is fundamentally distinct from the self-relation to one's internal zones of oblivion for which I argue later in this book. See especially chapter 5.

72. Farahany, *Battle for Your Brain*, 84

73. The subject of this sentence in the original is "la 'vie privée,'" which Richard Howard translates literally as "the 'private life,'" but which also refers to that which the French right to privacy protects. Since he and Warren and Brandeis and I are all talking about the same thing, I have decided to alter Howard's literal translation. The original reads: La «vie privée» n'est rien d'autre que cette zone d'espace, de temps, où je ne suis pas une image, un objet.

C'est mon droit politique d'être un sujet qu'il me faut défendre." Roland Barthes, *Camera Lucida: Reflections on Photography* (New York: Hill and Wang, 2010), 15; Roland Barthes, *La Chambre Claire: Note Sur La Photographie*, Cahiers Du Cinéma Gallimard (Paris: Gallimard, 1980), 32.

74. Barthes, *Camera Lucida*, 12.

75. Emerson, "Self Reliance."

76. Cf. Alison M. Jaggar, *Feminist Politics and Human Nature* (Totowa, NJ: Rowman and Allanheld, 1983).

77. Michel Foucault, "For an Ethics of Discomfort," in *Power*, ed. James D. Faubion, *The Essential Works of Foucault 1954-1984*, vol. 3 (New York: New Press, 2000), 444.

78. Foucault, "For an Ethics of Discomfort." This is Paul Rabinow's translation of this line, which appears in his introduction to the first volume of Foucault's collected works. Michel Foucault, *Ethics: Subjectivity and Truth*, vol. 1, *The Essential Works of Foucault 1954-1984* (New York: New Press, 1997), vxiii.

79. Warren and Brandeis, "Right to Privacy," 201.

2. Privacy, Perception, and Agency

1. Gay Talese, *The Voyeur's Motel* (New York: Grove Press, 2016), 26.

2. Steve Inskeep, "Glenn Greenwald: NSA Believes It Should Be Able to Monitor All Communication," radio, May 12, 2014, https://www.npr.org/sections/thetwo-way/2014/05/12/311619780/glenn-greenwald-nsa-believes-it-should-be-able-to-monitor-all-communication.

3. This is a definition of harm associated most prominently with Thomas Nagel and Joel Feinberg, who calls harms "setbacks to interests." Joel Feinberg, *The Moral Limits of the Criminal Law*, Volume 1, *Harm to Others* (New York: Oxford University Press, 1984), 31; Thomas Nagel, "Death," in *Mortal Questions* (Cambridge: Cambridge University Press, 2012), 1-10. That harms happen to one, rather than be brute facts about one's life, explains why although it might put one at a disadvantage to be born poor or a member of a disfavored group in a racist or homophobic society, we cannot say that one is harmed in being born one way rather than another.

4. A corollary of this observation will allow us to quickly dismiss one possible view of the harm that Ben suffers. It might be supposed that the harm Ben suffers is analogous to the harm of having his credit card number stolen but not used. He is in Foos's power, in a sense, which we might characterize as a harm in addition to the harm of that power's eventual use. But this still does not explain in what sense, particular to privacy, Ben would be in Foos's power—especially because there is simply no way that Foos could use the information he gleans to Ben's detriment.

5. A quick example of the token-type analytic for readers unfamiliar with this somewhat technical piece of lingo: dogs (type) are quadrupeds, but this one (token) has only three legs. However, it still shares whatever sufficient set of features with the type that permit us to identify it as a dog.

6. "We want to *do* certain things, and not just have the experience of doing them. In the case of certain experiences, it is only because first we want to do the actions that we want the

experiences of doing them or thinking we've done them." Robert Nozick, *Anarchy, State, and Utopia* (Malden, MA: Blackwell, 2012), 43. This is also known as the "objective view of well-being," but I opt for the language of externality for two reasons. First, to pick out the idea that events external to an individual's mental processes and relations with others can be said to affect his life. And second, to avoid getting drawn into metaethical controversies about the nature of value. I am grateful to Michael Rosen for several discussions on this point. In any case, the view of privacy I offer below does not depend on the truth of the externalist view of well-being, even if the ordinary view of *Voyeur* might.

7. Andrei Marmor thinks that we wouldn't have an interest in privacy against such aliens because, given the impossibility of contact, "we cannot have an interest in how we present ourselves to them." "How would our lives go less well, in any sense whatsoever" he asks rhetorically, "by knowing that the aliens are watching?" How indeed. Marmor's view of what privacy protects is of an interest in controlling how we appear to others, based in the value of such control for "shap[ing] the social life we want to have." *Voyeur* makes trouble for this class of views, which we will take up at the end of this chapter. Andrei Marmor, "What Is the Right to Privacy?," *Philosophy and Public Affairs* 43, no. 1 (2015): 11.

8. It is possible that *Y* even benefits *X* in this scenario—by blowing out a dangerous candle, say. By analogy we can imagine secret invasions of our privacy by a beneficent state or private actor and suppose that the constant snooping benefits us in some way like this, all without our ever knowing it. I think we would still find this sort of unobtrusive anti-private paternalism troubling—possibly even more so than simple spying. Why we should be troubled, especially if it appears that we stand to gain without suffering any apparent harm, is the question of this chapter. The harmless trespass example has its origins in Arthur Ripstein, "Beyond the Harm Principle," *Philosophy and Public Affairs* 34, no. 3 (2006): 215–245.

9. For a sampling of those views, see Ferdinand David Schoeman, ed., *Philosophical Dimensions of Privacy: An Anthology* (Cambridge: Cambridge University Press, 1984); Marmor, "What Is the Right to Privacy?"; Helen Nissenbaum, "Privacy as Contextual Integrity," *Washington Law Review* 79, no. 1 (2004): 119–157. This point also seems to vindicate Warren and Brandeis's frequently mocked assertion that the harms associated with privacy are not "material" in nature but "spiritual."

10. I borrow the term from James Rachels, *The End of Life: Euthanasia and Morality* (Oxford: Oxford University Press, 1986).

11. It is important to recall here that the biographical dimension of Ben's life is different from his reputation, and that a change in the former does not necessarily entail any effect on the latter. Obviously, Ben's reputation isn't affected by spying. Moreover, it is not uncommon that the reputations of individuals or groups are unaffected when contradictory information comes to light.

12. Rosenblum, *Another Liberalism: Romanticism and the Reconstruction of Liberal Thought* (Cambridge, MA: Harvard University Press, 1987), 50.

13. This idea will reappear in chapter 4's discussion of the rampant documentation of human life, where we will encounter Michel Foucault's genealogy and critique of "the constitution of an individual as a describable, analyzable object . . . to maintain him in his

individual features . . . under the gaze of a permanent corpus of knowledge." Michel Foucault, *Discipline and Punish: The Birth of the Prison* (New York: Vintage, 1995), 190.

14. To be sure, by smelling me you might learn that I smell bad, and I might not want you to learn that about me. The difference between this bit of information and the sorts of information that we typically transmit via sound and sight is that we do not tend to understand the bad smell to reflect on my personality—that is, on the part of my life that is up to me.

15. Aristotle, in the *Metaphysics*, claimed that "we prefer sight, generally speaking, to all the other senses. The reason for this is that of all the senses sight best helps us to know things, and reveals many distinctions" (*Metaphysics* 980a 21, from Aristotle, *Aristotle in 23 Volumes*, vols. 17 and 18, translated by Hugh Tredennick [Cambridge, MA, Harvard University Press]). It does so perhaps because of the largeness and richness of the visual field compared to others, but more importantly because it is the primary sense in which the perception of particular differences leads to the perception of general commonalities. In other words, the idea of differences is as meaningless as a language which had a distinct word for every separate object or appearance *unless* the concept of differences operates against a background of similarity. For instance, through sight we come to know that people can appear all sorts of different ways, but also that they're people, etc. Michel Foucault raises this interpretation in the first of his 1970 lectures at the Collège de France, titled in English Michel Foucault, *Lectures on the Will to Know*, ed. Arnold I. Davidson and trans. Graham Burchell (New York: Picador, 2015).

16. Under this description, we might understand the harm that Ben suffers in terms of dignity. Ben might have an interest in not being treated merely as a means for the satisfaction of voyeuristic desires; or he might have a dignity-type interest in not being thrust on stage, literally or metaphorically, independent of his awareness of what's happened to him. One could certainly make an argument along these lines, but I find this way of thinking overly vague and unsatisfying. For one thing, it doesn't need the specifics of the case—in fact, it need not have anything to do with privacy. This means that it won't be able to answer the question of this chapter: that is, why *Voyeur* should be an injury to Ben's dignity—if dignity is what interests us—while Foos observing Ben on the street is not. Dignity might be involved here, but it will not tell us what about being seen under certain circumstances gives us reason to think that it is.

Likewise, we might think that the reason Foos's spying is an affront to Ben's dignity is because the walls of a hotel room send an implicit message of non-consent to being seen (as being on the street doesn't). But that will still not explain why we should care about consent in the case of the hotel room and not in the case of the street (i.e., nobody thinks it violates Ben's privacy to see him on the street if he goes out but wishes everyone to look away from him). A view of the value of being unaccountable, to individuals and to society, permits us to make a distinction between the motel and the street, and to place moral weight on consent in the former but not the latter.

17. This is a pretty common view among philosophers of agency and self-knowledge. See, e.g., Frankfurt, "Freedom of the Will and the Concept of a Person"; Charles Taylor, "What Is Human Agency?," in *Human Agency and Language*, Philosophical Papers 1 (Cambridge:

Cambridge University Press, 1985), 15–44; Christine M. Korsgaard, *Self-Constitution: Agency, Identity, and Integrity* (Oxford: Oxford University Press, 2009).

18. I am not committed to a claim about whether this inference is justified or not, only that it's one we make all the time, and one we should expect others to make when we appear. Compare Gilbert Harman's discussion of the "fundamental attribution error" in Gilbert Harman, "Moral Philosophy Meets Social Psychology: Virtue Ethics and the Fundamental Attribution Error," *Proceedings of the Aristotelian Society* 99 (1999): 315–331.

19. E.g., Harry G. Frankfurt, "Freedom of the Will and the Concept of a Person," in *The Importance of What We Care about: Philosophical Essays* (Cambridge: Cambridge University Press, 1988), 11–25; Christine M. Korsgaard, *Self-Constitution: Agency, Identity, and Integrity* (Oxford: Oxford University Press, 2009); Charles Taylor, "What Is Human Agency?," in *Human Agency and Language*, Philosophical Papers 1 (Cambridge: Cambridge University Press, 1985), 15–44.

20. We needn't wade into debates about the metaphysics of subjectivity to notice that the common idea of getting out of one's head or loosening of one's grip on oneself does not depend upon or presume the existence of a transcendent self that gets out of or lets go of something, notwithstanding the grammatical implication. Indeed, in its ordinary meaning, the idiom of "getting out of one's head" or "letting oneself go" neither requires nor implies the existence of a transcendent self (or the Nietzschean idea that Judith Butler describes as "the metaphysics of substance" in *Gender Trouble: Feminism and the Subversion of Identity* [New York: Routledge Classics, 2006]). Rather, the ordinary meaning of those phrases is something closer to "stop insisting so much on being yourself" or "set aside the activity of being one way rather than another for a moment" lest that activity become a burden or force of deformation. As Robert Creeley puts it in his wonderful poem "The Rain," "What am I to myself / that must be remembered, / insisted upon so often? / [. . .] am I to be locked in this / final uneasiness." Creeley sets the experience of lying in bed and listening to the rain in opposition to the uneasy insistence of identity. I think most of us have probably lain in bed some night and listened to the rain. If you have, you may remember that as one gives oneself over to the sound of the rain one feels the grip on oneself slowly relax in a way that is healthful and restorative (and not dangerous, in the sense of the negative loss of self-accountability indicated by the injunction to "get a grip on yourself"). Robert Creeley, *The Collected Poems of Robert Creeley, 1945–1975* (Berkeley: University of California Press, 2006), 207.

21. For examples of this view from three philosophers with significantly different prior philosophical commitments, see Taylor, "What Is Human Agency?"; Frankfurt, "Freedom of the Will"; Korsgaard, *Self-Constitution*. As we will see in chapter 4, even a thoroughgoing critic of agency like Michel Foucault (or indeed, even Nietzsche) argued for the value of experiences and opportunities in which the self can be said to disintegrate or detach from itself ("*se déprendre de soi-même*"), albeit in the name of human well-being rather than the well-functioning of agency.

22. Taylor, "What Is Human Agency?," 41–42.

23. Taylor, "What Is Human Agency?," 42.

24. Taylor, "What Is Human Agency?," 42.

25. This is Harry Frankfurt's point about the centrality of caring and the impermanence and care-dependence of value. See, e.g., Harry G. Frankfurt, "The Importance of What We Care About," in *The Importance of What We Care about: Philosophical Essays* (Cambridge: Cambridge University Press, 1988), 80–94; Harry G. Frankfurt, *Taking Ourselves Seriously and Getting It Right*, The Tanner Lectures in Moral Philosophy (Stanford, CA: Stanford University Press, 2006).

26. Frankfurt, "What We Care About," 84.

27. "How tedious a person can be when striving to impress the world with personality." Yiyun Li, *Dear Friend, from My Life I Write to You in Your Life*, 1st ed. (New York: Random House, 2017), 28.

28. Harry G. Frankfurt, "Three Concepts of Free Action," in *The Importance of What We Care About: Philosophical Essays* (Cambridge: Cambridge University Press, 1988), 54.

29. Preface to Harry G. Frankfurt, *The Importance of What We Care about: Philosophical Essays* (Cambridge: Cambridge University Press, 1988), xi.

30. Frankfurt, *Taking Ourselves Seriously*, 2.

31. To be sure, when it comes to experiencing works of art, we are often in public, although we should also note that we tend to describe profound experiences in this area with analogies to privacy: "the world fell away" or "I felt like I was the only one in the room and she was singing just to me." The erotic experiences of physical intimacy come closer, I think, to a view of what privacy would have given Ben in *Voyeur*. The thought here is that some quality of the erotic experience would fail to obtain if those aliens were watching. And it would fail to obtain not for reasons of shame, but for reasons that have to do with the way of relating to the other in the open space of the erotic. This is why you might not want your greatest friend and confidant to watch your lovemaking, even if he or she has already heard all the juicy details from you.

32. John Keats, Letter to George and Tom Keats, December 22, 1818. Compare this gloss of Fitzgerald's, which connects the idea to agency: "The test of a first-rate intelligence is the ability to hold two opposed ideas in the mind at the same time, and still retain the ability to function." F. Scott Fitzgerald, "The Crack-Up," in *The Crack-Up* (New York: New Directions, 1993), 69. John Keats, *Selected Letters*, Oxford World's Classics (Oxford: Oxford University Press, 2009).

33. Note again the disconnect with touch, smell, and taste: it is hard to imagine that these ineffable areas could be given expression, either intentionally or by accident, in any of these sensual domains. My view here also looks a bit like translating Arendt's ideas about the oblivion of privacy from the domain of social ontology to that of self-relation. And since we will take up Erving Goffman in later chapters, it is worth pointing out that this view is at odds with his distinction between frontstage and backstage. What gives one a sense of depth, meaning, and fundamental independence from instrumentality are those parts of one's own life that are obscure even to oneself, and not merely hidden. Erving Goffman, *The Presentation of Self in Everyday Life* (New York: Doubleday, 1990).

34. We might well say that Foos has altered Ben's secrecy in one or more ways, but he certainly has not destroyed it.

35. See, e.g., Judith Wagner DeCew, "The Feminist Critique of Privacy: Past Arguments and New Social Understandings" in Beate Roessler and Dorota Mokrosinska, eds., *Social*

Dimensions of Privacy: Interdisciplinary Perspectives (Cambridge: Cambridge University Press, 2015), 85–103; Jean Bethke Elshtain, *Public Man, Private Woman Women in Social and Political Thought: Second Edition* (Princeton, NJ: Princeton University Press, 1993); Anita L. Allen, *Uneasy Access: Privacy for Women in a Free Society* (Totowa, NJ: Rowman and Littlefield, 1987).

For a clear distinction between the various usages of the "private" that draws a conceptual contrast between the liberal private sphere and the feminist concern with the private as the domestic, see Jeff Weintraub's contribution in, but also generally, Jeff Alan Weintraub and Krishan Kumar, eds., *Public and Private in Thought and Practice: Perspectives on a Grand Dichotomy* (Chicago: University of Chicago Press, 1997).

36. The most famous of these was probably Catharine A. MacKinnon, *Toward a Feminist Theory of the State* (Cambridge, MA: Harvard University Press, 1989), 191. Judith Wagner DeCew offers an elegant critique of MacKinnon's argument as either confusing privacy with the liberal private sphere or attacking the strawman I describe in a successive paragraph. In any event, MacKinnon's argument is one rarely heard today, in the age of surveillance capitalism, when the structures of power and domination tend to be seen as having an interest in the domestic sphere's transparency rather than its obscurity. DeCew, "Feminist Critique of Privacy."

37. Seyla Benhabib, *The Reluctant Modernism of Hannah Arendt* (Thousand Oaks, CA: Sage, 1996), 214.

38. Sarah Elizabeth Igo, *The Known Citizen: A History of Privacy in Modern America* (Cambridge, MA: Harvard University Press, 2018), 117. Betty Friedan, *The Feminine Mystique* (New York: Norton, 2001).

39. Simone Browne, *Dark Matters: On the Surveillance of Blackness* (Durham, NC: Duke University Press, 2015).

40. Édouard Glissant, *Poetics of Relation*, trans. Betsy Wing (Ann Arbor: University of Michigan Press, 1997), 194.

41. Rachels, "Why Privacy Is Important," 330. Jeffrey Reiman offers a different, more devastating critique of this view of information and intimacy in "Privacy, Intimacy, and Personhood," *Philosophy and Public Affairs* 6, no. 1 (1976): 26.

42. The kind of knowledge we acquire about another or ourselves by this experience (if any) is not the sort that can take the form of facts or information, but rather a kind of familiarity that resists the translation into information. Philosophers call this type of knowledge "knowledge by acquaintance" to distinguish it from the "knowledge that" of information, which they call "propositional knowledge." The awkward terminology is necessary in English but not in many other languages, which have two words for the different types of knowledge: for instance, *conocer, connaître,* and *kennen* for knowledge by acquaintance, and *saber, savoir,* and *wissen* for propositional knowledge.

43. Pablo Neruda, "Oda al Libro (I)" (translation mine).

44. The analogy "description is to certain forms of experience as pornography is to sex" I take from Blake Smith, "Against Description," *Substack* (newsletter), July 7, 2023 (available at https://blakeesmith.substack.com/p/against-description?utm_source=post-email-title &publication_id=1186693&post_id=133570477&isFreemail=true&utm_medium=email).

3. Hiding in Private

1. See, e.g., Sherry Turkle, "Always-on/Always-on-you: The Tethered self," in *Handbook of Mobile Communication Studies*, ed. James E. Katz (Cambridge, MA: MIT Press, 2008).

2. Matthew B. Crawford, *The World beyond Your Head: On Becoming an Individual in an Age of Distraction* (New York: Farrar, Straus and Giroux, 2015).

3. Bernard E. Harcourt, *Exposed: Desire and Disobedience in the Digital Age* (Cambridge, MA: Harvard University Press, 2015), 25.

4. Justin E. H. Smith, *The Internet Is Not What You Think It Is: A History, a Philosophy, a Warning* (Princeton, NJ: Princeton University Press, 2022), 11.

5. "'Is There Any Privacy?,'" *Hartford Daily Courant*, October 3, 1874.

6. Sesame and Lilies (1865), quoted in Richard Sennett, *The Conscience of the Eye: The Design and Social Life of Cities*, 1st ed. (New York: Knopf, 1990), 20.

7. "Mind Your Own Business: Rev. George C. Lorimer Says This Is an Age without Privacy," *Boston Daily Globe*, July 11, 1892, p. 2.

8. Henry James, *The Notebooks of Henry James*, ed. Francis O. Matthiessen (Chicago: University of Chicago Press 1981), 82 (emphasis in the original).

9. Warren and Brandeis, "The Right to Privacy," *Harvard Law Review* 4, no. 5 (Dec. 15, 1890), 196.

10. Wallace Stevens, "The Noble Rider and the Sound of Words," in *The Necessary Angel* (New York: Vintage, 1951), 18, 20. As with the moral umbrage surrounding newspapers before and television afterward, Stevens's complaint was characteristic of the feeling of the time. For instance, Marshall McLuhan writes: "Radio provides a speed-up of information that also causes acceleration in other media. It certainly contracts the world to village size, and creates insatiable village tastes for gossip, rumor and personal malice." (Marshall McLuhan, *Understanding Media: The Extensions of Man* [Cambridge, MA: MIT Press, 1995], 306.) Stevens and McLuhan (and Stein in the following note) echo the older complaint about the spread of publicity that has been with us since the origins of modern privacy in the nineteenth century.

11. Gertrude Stein, "Reflection on the Atomic Bomb," in *Gertrude Stein: Writings 1932–1946*, ed. Catharine R. Stimpson and Harriet Chessman (New York: Library of America, 1998), 823.

12. Günther Anders, "Die Antiquiertheir des Menschen" (1956), quoted in Peter Weibel, "Pleasure and the Panoptic Principle," in *CTRL [Space]: Rhetorics of Surveillance from Bentham to Big Brother* (Cambridge, MA: MIT Press, 2002), 210.

13. Harcourt, *Exposed*.

14. Arendt, *The Origins of Totalitarianism* (New York: Harcourt, 1994), 478.

15. In just a moment I will have more to say about the failures of these views and their general approach to the subject, but first let us linger one minute longer with the sense of perversity that comes from describing the Franks' condition in the secret annex in terms of privacy rather than hiding. The few approaches to the question of privacy that consider hiding (of which I am aware) beg the question of its identity with privacy; they determine that privacy is a form of concealment with regard to perception of oneself or some object,

see that their description could also describe hiding, and therefore conclude that the two must be the same. I do not want to give the impression that I, too, assume the truth of what I seek to demonstrate by assuming without justification that privacy and hiding are in fact distinct. For one thing, this chapter argues at length for the validity of this distinction. Moreover, the sense of perversity or moral transgression attendant upon the use of privacy to describe the Franks' case, and the long history and common sense about nonidentical usage, give reason to start from the defeasible presumption that hiding and privacy are not only distinct forms of concealment, but that we might have reason not to want what once was private to become more like hiding. This presumption is exceedingly strong, I think. An argument, even a very strong one, which held privacy and hiding were identical would require, among other things, giving a special, technical meaning to "hiding" that differs from and in many cases opposes the ordinary usage of the word. And then, I think, we would be talking about something else altogether.

16. For a thorough rundown of the control-access debate, see Jakob Mainz, "An Indirect Argument for the Access Theory of Privacy," *Res Publica* 27, no. 3, 309–328 (2021).

17. These come from Mainz, "An Indirect Argument for the Access Theory of Privacy," who either quotes these formulae from other philosophers or composes them himself with the aim of synthesizing the views of a very wide range of philosophers. For a wealth of references to the multitude of views that Mainz distills into these abstract descriptions, see Mainz's article.

18. Even the more properly "descriptive" disciplines of sociology and anthropology would regard the descriptions above as unacceptably thin, telling us little about the actual practice and values of privacy to which they purport to refer (if, however, they do reveal a great deal about the cultures of contemporary analytic philosophy and political theory).

19. I take this to be a serious challenge to a wide range of contemporary views on privacy, but since the point of this book is not to disprove their views but to elaborate my own, I leave this challenge to the side for now. Many, though not all, "normative" theories of privacy also fail to distinguish between hiding and privacy. However, for present purposes I leave these theories aside, since my argument isn't just that theories of privacy fail because they don't distinguish between hiding and privacy, but that this failure points us to the elements that a fuller theory of privacy ought to have.

20. Luke 8:17 (KJV); see also Matthew 10:26 ("For there is nothing covered that will not be revealed, and hidden that will not be known").

21. William Shakespeare, *Merchant of Venice*, Act II, Scene 2, 643–645.

22. Sigmund Freud and James Strachey, *Introductory Lectures on Psychoanalysis* (New York: Norton, 1977). Notice that here the hidden and the secret are more or less interchangeable. I think this is true, as was once reflected by the now antiquated use of "to secret" for "to hide." In modern usage, especially in the era of writing, hiding has tended to refer to physical objects, and secrets to information.

23. Here we find another instance of moral and political reasons, in addition to conceptual ones, to distinguish hiding from privacy and to embrace the value of oblivion. Presumptions about who is hiding something and who is merely private or reticent—and cultural assumptions about what those who are presumed to be hiding something are thereby

concealing—can be powerful forces for the perpetuation of injustices and the reinforce-ment of inegalitarian social structures. In a racist society, for instance, where one group is throught to be devious or particularly dangerous, a lack of personal transparency by a member of that group can be coded as a threat. Those of us who lived through the aftermath of the 9/11 terrorist attacks saw firsthand how the weight of "See Something, Say Some-thing" and "Privacy is for those who have something to hide" fell with powerful inequality upon different groups of citizens. Or think of how young Black men in the United States (and elsewhere) have been portrayed for generations as dangerous, which leads police and others to presume that they are likely hiding something perhaps dangerous or illegal—or at least that they are more likely to be hiding something than, say, young white men of similar age and circumstances, and that whatever it is they are hiding is more likely to pose a threat. In turn, they are pressured to embody an elevated degree of personal transparency in public and especially in their relations with the authorities. This is part of "the talk" that Black parents often have with their children about interactions with the police. These young men, like many others, are being deprived of the good of personal oblivion while out in the world, which their white counterparts still enjoy. This is another failure of justice with regard to the distribution of oblivion in society. Obviously there are many other (and more im-portant) injustices faced by marginalized communities, but it is worth noticing the unjust distribution of this particular good, in addition to the many others that are unfairly denied to citizens in racist, sexist, and homophobic societies.

24. You might say that an object or person could "come to be hidden" by accident, in the way that a coin I drop into a pile of leaves will be hidden by it (or how wandering in the woods I might be hidden from general view). But you cannot say that I have hidden the coin, or that I am hiding in the forest, in the active sense I am trying to pick out here unless I in-tended to do so.

25. Anne Frank, *The Diary of a Young Girl: The Definitive Edition*, ed. Otto Frank and Mirjam Pressler, trans. Susan Massotty (New York: Bantam, 1997), 53.

26. Stanley Cavell, *Must We Mean What We Say?*, rev. ed. (Cambridge, UK : Cambridge University Press, 2002), 261.

27. Koch, *Solitude: A Philosophical Encounter* (Chicago: Open Court, 1994), 27.

28. Arendt, *Origins of Totalitarianism*, 476.

29. The US American archetype of this view is, of course, Henry David Thoreau. See, Thoreau, *A Week on the Concord and Merrimack Rivers; Walden, or, Life in the Woods; The Maine Woods; Cape Cod.*

30. Barthes, *Camera Lucida: Reflections on Photography* (New York: Hill and Wang, 2010), 15.

31. The connection hinted at here between trust, privacy, and oblivion will be taken up at length in chapter 5.

32. Kevin Everod Quashie, *The Sovereignty of Quiet: Beyond Resistance in Black Culture* (New Brunswick, NJ: Rutgers University Press, 2012), 15.

33. Quashie, *Sovereignty of Quiet*, 22.

34. See, e.g., Thomas Nagel, "Concealment and Exposure," in *Concealment and Expo-sure: And Other Essays* (Oxford: Oxford University Press, 2002), 3–26; Anita L. Allen,

Unpopular Privacy: What Must We Hide? (Oxford: Oxford University Press, 2011); Michael Warner, *The Trouble with Normal: Sex, Politics and the Ethics of Queer Life* (New York, NY: Free Press, 1999); Jill Locke, *Democracy and the Death of Shame: Political Equality and Social Disturbance* (Cambridge: Cambridge University Press, 2017).

35. Turkle, "Always-on/Always-on-You."

36. Sherry Turkle, *Alone Together: Why We Expect More from Technology and Less from Each Other* (New York: Basic Books, 2011), 174.

37. Turkle, *Alone Together*, 174; Donna Freitas, *The Happiness Effect: How Social Media Is Driving a Generation to Appear Perfect at Any Cost* (New York: Oxford University Press, 2019), 218; Turkle, *Alone Together*, 172.

38. Turkle, *Alone Together*, 163.

39. Pavesich v. New England Life Ins. Co.—122 Ga. 190, 50 S.E. 68 at 220.

40. Freitas, *Happiness Effect*, 23.

41. Two sides of a single coin: the person who checks their hiding spot to make sure an item is still there, and the person who checks their post on social media to see if it has garnered any "engagement."

42. Freitas, *Happiness Effect*, 18.

43. Freitas, *Happiness Effect*, 73.

44. In a poll of 727 students, 79 percent agreed that "I'm aware that my name is a brand and I need to cultivate it carefully." Freitas, *Happiness Effect*, 80.

45. Turkle also notes the obverse phenomenon, in which people gathered in public are all lost in their cell phones. She describes a train station drained of its character as a public space because everyone who is physically present there is also "somewhere else," immersed in their own isolated, personal worlds of conversation and scrolling. Turkle, "Always-on/Always-on-you," 122.

46. Turkle, *Alone Together*, 259.

47. Freitas, *Happiness Effect*, 46.

48. My thanks to Jennifer Forestal for raising this objection.

49. Harcourt, *Exposed*, 26.

50. Jeffrey Reiman, "Privacy, Intimacy, and Personhood," *Philosophy and Public Affairs* 6, no. 1 (1976): 39. Note that Reiman's description of what privacy is does not fall prey to the same mistakes as the so-called "descriptive accounts." Reiman's view that "privacy is a social ritual by means of which an individual's moral title to his existence is conferred" (39) includes a normative element because it is the description of a normative practice. It is not necessarily normative in the way we typically use that term to describe a theory as advocating for a particular way of structuring human life, the avoidance of which is, one imagines, what attracts one to come up with a "merely descriptive account" in the first place. The accuracy of Reiman's description does not depend upon any prior moral commitments. Nor does it require that we accept that it's a good thing for people to have moral title to their existence conferred to them (if you think this bad, then Reiman's argument gives you reason to not want the social ritual of privacy in your society).

51. Reiman, "Privacy, Intimacy, and Personhood," 39, 42–43. If this is the ur-expereince of privacy, as I think it must be, then we should expect the image of the individual alone in her room to be central to our thinking on the subject.

52. Freitas, *Happiness Effect*, 209.

53. Hannah Arendt, "The Crisis in Education," in *Between Past and Future: Eight Exercises in Political Thought* (New York: Penguin, 2006), 183.

54. Hannah Arendt, *The Human Condition*, 2nd ed. (Chicago: University of Chicago Press, 1998).

55. Arendt, *Origins of Totalitarianism*, 474.

56. William Wordsworth, "The World Is Too Much with Us," originally published in *Poems, in Two Volumes* (1807).

57. Arendt, *Origins of Totalitarianism*, 476.

58. Arendt, *Origins of Totalitarianism*, 477.

59. Arendt, *Origins of Totalitarianism*, 478.

4. Memory and Oblivion

1. Oliver Wendell Holmes, "The Stereoscope and the Stereograph," in *Classic Essays on Photography* (New Haven, CT: Leete's Island Books, 1980), 81.

2. Matthew 10:30 (New International Version Bible).

3. See, e.g., Gabriel Josipovici, *Forgetting* (Manchester, UK: Little Island Press in collaboration with Carcanet, 2020); Rima Basu, "The Importance of Forgetting," *Episteme* 19, no. 4 (2022): 471–490; Guilherme Cintra Guimarães, "Privacy, Social Memory and Global Data Flows," in *Global Technology and Legal Theory: Transnational Constitutionalism, Google and the European Union* (London: Routledge, 2019), 113–165; Elena Esposito, "Algorithmic Memory and the Right to Be Forgotten on the Web," *Big Data and Society* 4, no. 1 (2017): 1–11; Jeffrey Rosen, "The Web Means the End of Forgetting," *New York Times*, July 21, 2010, https://www.nytimes.com/2010/07/25/magazine/25privacy-t2.html; Viktor Mayer-Schönberger, *Delete: The Virtue of Forgetting in the Digital Age* (Princeton, NJ: Princeton University Press, 2009); Noam Tirosh, "Reconsidering the 'Right to Be Forgotten': Memory Rights and the Right to Memory in the New Media Era," *Media, Culture and Society* 39, no. 5 (July 2017): 644–660; Jonah Bossewitch and Aram Sinnreich, "The End of Forgetting: Strategic Agency beyond the Panopticon," *New Media and Society* 15, no. 2 (2013): 224–242; Mariano Sigman, "Bridging Psychology and Mathematics: Can the Brain Understand the Brain?," *PLOS Biology* 2, no. 9 (September 14, 2004): e297; Scott A. Small, *Forgetting* (New York: Crown, 2021); Yair Neuman, "Context and Memory: A Lesson from Funes the Memorious," in *Studies in Multidisciplinarity*, ed. Laura A. McNamara et al., vol. 6 (Elsevier, 2008), 229–238; Stevan Hardna, "To Cognize Is to Categorize: Cognition Is Categorization," in *Handbook of Categorization in Cognitive Science*, ed. Henri Cohen and Claire Lefebvre (Oxford: Elsevier Science Ltd., 2005), 19–43; Geoffrey C. Bowker, *Memory Practices in the Sciences* (Cambridge, MA: MIT Press, 2005); Luis Fornazzari et al., "Hyper Memory, Synaesthesia, Savants Luria and Borges Revisited," *Dementia and Neuropsychologia* 12, no. 2 (2018): 101–4; N. Ohri, A. Gill, and M. Saini, "Borges and the Art of Forgetting," *European Psychiatry* 64, no. S1 (2021): S753.

4. For instance, this article published in a medical journal compares the case of Funes side by side with that of an actual mnemonist studied by neuropsychologist A. R. Luria: "In this paper, we investigated two subjects with superior memory, or hyper memory: Solomon

Shereshevsky, who was followed clinically for years by A. R. Luria, and Funes the Memorious, a fictional character created by J. L. Borges." Fornazzari et al., "Hyper Memory, Synaesthesia, Savants Luria and Borges Revisited," 101. See also Basu: "This short story points to an important lesson about forgetting: 'without forgetting, the human species would have to relive the past continuously and never live in the present moment. Without forgetting, there would be no future'" ("Importance of Forgetting," 472).

5. Part of the reason for this, as we will show below, is that Funes's type of memory, which likely would be intolerable, is almost certainly impossible for a nonfictional human.

6. Small, *Forgetting*, 7.

7. From Jorge Luis Borges, "Funes, His Memory," in *Collected Fictions*, trans. Andrew Hurley (New York: Penguin, 1998), 135. I have altered Andrew Hurley's translation of the story where necessary to express more clearly, if perhaps less mellifluously, the meaning of Borges's text.

8. Borges, "Funes, His Memory," 135.

9. See, for instance, Friedrich Nietzsche, *Unfashionable Observations*, trans. Richard T. Gray, vol. 2 of *The Complete Works of Friedrich Nietzsche* (Stanford, CA: Stanford University Press, 1995); Friedrich Nietzsche, *On the Genealogy of Morality*, trans. Maudemarie Clark and Alan J. Swensen (Indianapolis, IN: Hackett, 1998). William James wrote: "If we remembered everything, we should on most occasions be as ill off as if we remembered nothing. . . . 'The paradoxical result [is] that one condition of remembering is that we should forget. Without totally forgetting a prodigious number of states of consciousness, and momentarily forgetting a large number, we could not remember at all.'" (James is quoting here from Théodule-Armand Ribot, *Les maladies de la mémoire* [Paris: Librarrie Germer Ballière, 1881]. William James 1842–1910, *The Principles of Psychology*, vols. 1–2 (New York: Dover, 1950), 680–681.)

10. Small, *Forgetting*, 7.

11. Mayer-Schönberger, *Delete*, 12.

12. Borges, "Funes, His Memory," 136. Careful readers will note that Funes is said to have been "*virtually* incapable of general, *Platonic* ideas" (emphasis mine). The modifiers "virtually" and "Platonic" cannot be ignored. Not only would Borges have been familiar with Plato's epistemology and metaphysics, but his prose is characterized, among other things, by a high degree of emphasis on precision in language used by narrators whose grasp or memory of the situation is imprecise or not fully known.

13. Hurley translates this as "To sleep is to take one's mind off the world," which also makes my point. *Distraerse* means literally "to take oneself away from" (from the world, in this case) but whose most common usage is equivalent to our similarly derived "to distract."

14. Borges, "Funes, His Memory," 136–137. Andrew Hurley translates "la presión de una realidad tan infatigable como la que día y noche convergía sobre el infeliz Irneo" as "pressure of a reality as inexhaustable as that which battered Ireneo." The original Spanish translates as "tireless," for which the English "inexhaustible" is one suitable synonym. However, the other sense of "inexhaustible" as "incapable of being depleted" fatally misses the point of Borges's Spanish here, which is that what exhausted Funes was his perfect accounting of the world around and within him.

15. Wallace Stevens, "The Noble Rider and the Sound of Words," in *The Necessary Angel* (New York: Vintage, 1951), 20.

16. Borges, "Funes, His Memory," 136–137.

17. The hubris of exhaustive knowledge is another theme in Borges; see, e.g., "On Exactitude in Science," in which an empire's cartographers aspire to create a 1:1 map of their territory.

18. Nietzsche, *Unfashionable Observations*, 90.

19. Josipovici, *Forgetting*, 33.

20. David Rieff, *In Praise of Forgetting: Historical Memory and Its Ironies* (New Haven, CT: Yale University Press, 2016), 129.

21. Nietzsche, *On the Genealogy of Morality*, 35.

22. Nietzsche, *Unfashionable Observations*, 87.

23. This is Google's official "Delisting Policy," which quotes directly from the European Court of Justice's landmark decision recognizing a right to be forgotten (Google Transparency Report, "Requests to Delist Content under European Privacy Law," available at: https://transparencyreport.google.com/eu-privacy/overview?hl=en). Some of this chapter's discussion of the right to be forgotten is adapted from Lowry Pressly, "The Right to be Forgotten and the Value of an Open Future," *Ethics* 135, no. 1 (October 2024). There I provide a more thoroughgoing (and technical) account of the relation between healthy agency and the possibility of forming relations un-preconditioned by knowledge of one's past.

24. Angelo Maietta, "The Right to Be Forgotten," *RECHTD. Revista de Estudos Constitucionais, Hermenêutica e Teoria Do Direito* 12, no. 2 (2020): 208.

25. Pedro Anguita Ramírez, "The Right to Be Forgotten in Chile: Doctrine and Jurisprudence," E-conférence, droit à l'oubli en Europe et au-delà (2017), 1.

26. Leticia Bode and Meg Leta Jones, "Ready to Forget: American Attitudes toward the Right to Be Forgotten," *Information Society* 33, no. 2 (March 2017): 76–85; Dawn Carla Nunziato, "The Fourth Year of Forgetting: The Troubling Expansion of the Right to Be Forgotten," *University of Pennsylvania Journal of International Law* 39, no. 4 (2018): 1011–1064; Indranath Gupta and Paarth Naithani, "Right to Be Forgotten in Case of Search Engines: Emerging Trends in India as Compared to the EU," *Journal of Data Protection and Privacy* 5, no. 3 (2022): 297–309.

27. For example, European Commissioner for Information Society and Media Viviane Reding characterized the right to be forgotten in Europe as a protection for "the basic right of privacy" in a *New York Times* op-ed. Viviane Reding, "Protecting Europe's Privacy," *New York Times*, June 17, 2013. The conflation of rights to privacy and to being forgotten is everywhere in scholarship and popular media.

28. It is practically axiomatic in the philosophy of privacy that a "violation of privacy is about ways in which information is obtained, and not about the content of the information per se." Andrei Marmor, "What Is the Right to Privacy?," *Philosophy and Public Affairs* 43, no. 1 (2015): 4.

29. Order of 6 November 2019—1 BvR 16/13 ("Right to be Forgotten I") (Federal Constitutional Court of Germany November 6, 2019); BBC, "German Murderer Wins 'Right to Be Forgotten'" November 27, 2019 (available at https://www.bbc.com/news/world-europe -50579297). Michael Bender, "Yachting and Madness," *Journal for Maritime Research* 15,

no. 1 (2013): 83–93. The case of the Apollonia is discussed both by Peter Noble and Ros Hogbin in *The Mind of the Sailor: An Exploration of the Human Stories behind Adventures and Misadventures at Sea* (New York: McGraw-Hill, 2001); and Michael Stadler, *Psychology of Sailing: The Sea's Effects on Mind and Body* (Camden, NJ: International Marine Publishing, 1988). See also Nic Compton, *Off the Deep End: A History of Madness at Sea* (London: Adlard Coles Nautical, 2017).

30. ABC News, "German Murderer Has 'Right to Be Forgotten' by Internet, according to Top Court," November 27, 2019, https://www.abc.net.au/news/2019-11-28/german-murderer -has-right-to-be-forgotten-constitutional-court/11745376.6.

31. "Right to be Forgotten I," 21 (emphasis supplied); BC, "German Murderer Wins 'Right to Be Forgotten,'" November 27, 2019 (available at https://www.bbc.com/news/world-europe -50579297).

32. For instance, recall Pavesich v. New England Life Insurance Company:

> The knowledge that one's features and form are being used for such a purpose and displayed in such places as such advertisements are often liable to be found brings not only the person of an extremely sensitive nature, but even the individual of ordinary sensibility, to a realization that his liberty has been taken away from him, and, as long as the advertiser uses him for these purposes, he can not be otherwise than conscious of the fact that he is, for the time being, under the control of another, that he is no longer free, and that he is in reality a slave without hope of freedom, held to service by a merciless master; and if a man of true instincts, or even of ordinary sensibilities, no one can be more conscious of his complete enthrallment than he is (220).

33. Viktor Mayer-Schönberger, *Delete: The Virtue of Forgetting in the Digital Age* (Princeton, NJ: Princeton University Press, 2009), 125.

34. Meg Leta Jones, *Ctrl + Z: The Right to Be Forgotten* (New York: NYU Press, 2016), 2, 93, 114.

35. Daniel J Solove, *The Future of Reputation: Gossip, Rumor, and Privacy on the Internet* (New Haven, CT: Yale University Press, 2008), 73.

36. Basu, "Importance of Forgetting," 473. For other examples of the common characterization that the right to be forgotten responds to the "perfect" or "total" memory of the Internet and its elimination of forgetting, see, e.g., Jonah Bossewitch and Aram Sinnreich, "The End of Forgetting: Strategic Agency beyond the Panopticon," *New Media and Society* 15, no. 2 (2013): 224–242; Kate Eichhorn, *The End of Forgetting: Growing up with Social Media* (Cambridge, MA: Harvard University Press, 2019); Elena Esposito, "Algorithmic Memory and the Right to Be Forgotten on the Web," *Big Data and Society* 4, no. 1 (2017): 1–11; Byron Reese, "Losing Our Ability to Forget," *Journal of Information Ethics* 22, no. 2 (Fall 2013): 5–8; Solove, *Future of Reputation*, 8; Jean-François Blanchette and Deborah G. Johnson, "Data Retention and the Panoptic Society: The Social Benefits of Forgetfulness," *Information Society* 18, no. 1 (2002): 36; Mónica Correia, Guilhermina Rêgo, and Rui Nunes, "Gender Transition: Is There a Right to Be Forgotten?," *Health Care Analysis* 29, no. 4 (2021): 283–300; Ludo Gorzeman and Paulan Korenhof, "Escaping the Panopticon over Time," *Philosophy and Technology* 30 (2017), 73–92, at 74; Jill Lepore, "The Cobweb: Can the Internet Be Archived?" *New Yorker*

(January 26, 2015); Cécile Terwangne, "The Right to Be Forgotten and Informational Autonomy in the Digital Environment," in *The Ethics of Memory in a Digital Age: Interrogating the Right to Be Forgotten* (Houndmills, UK: Palgrave Macmillan, 2014), 87.

37. Rosen, "Web Means the End." Recall Matthew Crawford's complaint about attention from the previous chapter, almost identical: "As our mental lives become more fragmented, what is at stake often seems to be nothing less than the question of whether one can maintain a coherent self."

38. Jean-François Blanchette and Deborah G Johnson, "Data Retention and the Panoptic Society: The Social Benefits of Forgetfulness," *Information Society* 18, no. 1 (2002): 33–45.

39. Socrates makes a similar point in his critique of writing in Plato's *Phaedrus* (274c–277a), although he goes even further to see external supplements to memory as actually undermining the faculty of memory. Plato, *Phaedrus*, trans. Alexander Nehamas and Paul Woodruff (Indianapolis, IN: Hackett, 1995).

40. For K, who went to prison in the 1980s, the disconnection of his name from the articles about his conviction would probably have the effect of restoring total obliviousness about his past, since it's hard to imagine that he generated much of an online presence while in prison except as relates to his convictions and crimes.

41. This coincides with one hypothesis about why sex-offender registries don't seem to decrease recidivism by previously convicted offenders: If offenders lose confidence that they could ever form new relations that are not preconditioned by accurate knowledge of their crimes, or that it is worth the trouble to change if they'll always be known as a sex-offender, then they may come to doubt whether it is worth it to reform. See, e.g., J. Prescott and Jonah E. Rockoff, "Do Sex Offender Registration and Notification Laws Affect Criminal Behavior?" *Journal of Law and Economics* 54, no. 1 (2011): 161–206; Kristen M. Zgoba, Wesley G. Jennings, and Laura M. Salerno, "Megan's Law 20 Years Later: An Empirical Analysis and Policy Review," *Criminal Justice and Behavior* 45, no. 7 (2018): 1028–1046.

42. Michel Foucault and Rux Martin, "Truth, Power, Self," in *Technologies of the Self: A Seminar with Michel Foucault* (Amherst: University of Massachusetts Press, 1988), 9.

43. I think there is, in fact, a deep symmetry between the two cases, which brings out a counterintuitive aspect of how we value memory. Funes's view of his own life in retrospection is a view of a life without possibility, without the fluidity or flux that ordinary, living memory has. Our memories are not fixed like a photograph but, as experience and science confirms, are a constant shifting process of rewriting, revision, introjection, and error. Although we sometimes lament losses or inaccuracies of memory, Borges's fable and the right to be forgotten reveal that our relationship to ourselves as creatures capable of change, surprise, and depth depends on a certain fluidity or encounter with the potentiality at the heart of personality. We will have more to say about this in chapter 5, so for now let us just recognize that Funes's case reflects this in the realm of value as K's does in the realm of society and politics. Although our lives cannot be Funes's life, the idea that human life, and our relationship to others and ourselves, is diminished when overwhelmed by information offers insight into the history of privacy (as seen in the next section), new focal points for our moral and political attention, and normative guidance for the practical structuring of society.

44. The connection of branding to the phenomenology of being shackled to one's past is sometimes hinted at, especially by reference to one's past as a "scarlet letter." For instance, Rosen writes: "The permanent memory bank of the Web increasingly means there are no second chances—no opportunities to escape a scarlet letter in your digital past" (Rosen, "Web Means the End").

45. Michel Foucault, *Discipline and Punish: The Birth of the Prison* (New York: Vintage, 1995), 172.

46. Foucault, *Discipline and Punish*, 209.

47. *Pavesich v. New England Life Ins. Co.*—122 Ga. 190, 50 S.E. 68 (1905).

48. Bert-Jaap Koops, "Forgetting Footprints, Shunning Shadows: A Critical Analysis of the Right to Be Forgotten in Big Data Practice," *SCRIPTed: A Journal of Law, Technology and Society* 8 (2011): 2. Note that this discussion of fixity and fluidity, and their attendant effects, takes place in the register of social ontology—not ontology proper or metaphysics—which means that if these changes are broadly understood to have occurred, they effectively have. There is no fluidity or fixity of the self or a photograph independent of our understanding that it is so, in other words.

49. Sarah Elizabeth Igo, *The Known Citizen: A History of Privacy in Modern America* (Cambridge, MA: Harvard University Press, 2018), 8–9.

50. Igo, *Known Citizen*; Craig Robertson, *The Passport in America: The History of a Document* (Oxford: Oxford University Press, 2010).

51. Igo, *Known Citizen*, 56 (citing Robert L. Floyd, "Privacy," Chicago Daily Tribune, September 27, 1925, p. 8).

52. Bernard E. Harcourt, *Exposed: Desire and Disobedience in the Digital Age* (Cambridge, MA: Harvard University Press, 2015), 176.

53. Igo, *Known Citizen*, 243.

54. See, e.g., Pete Grieve, "To Save Money on Insurance, Drivers Are Agreeing to 'Incredibly Intrusive' Monitoring Technology," *Money*, accessed July 30, 2023, https://money.com /usage-based-car-insurance-gaining-customers/; Randy Bean, "Transforming the Insurance Industry with Big Data, Machine Learning and AI," *Forbes*, accessed July 30, 2023, https://www.forbes.com/sites/randybean/2021/07/06/transforming-the-insurance-industry -with-big-data-machine-learning-and-ai/.

55. See, e.g., Craig Timberg and Isaac Stanley-Becker, "Cambridge Analytica Database Identified Black Voters as Ripe for 'Deterrence,' British Broadcaster Says," *Washington Post*, September 28, 2020, https://www.washingtonpost.com/technology/2020/09/28/trump-2016 -cambridge-analytica-suppression/.

56. Latanya Sweeney, Akua Abu, and Julia Winn, "Identifying Participants in the Personal Genome Project by Name (A Re-Identification Experiment)," arXic:13.04.7605 (2013) (available at https://arxiv.org/abs/1304.7605); Latanya Sweeney et al., "Re-Identification Risks in HIPAA Safe Harbor Data: A Study of Data from One Environmental Health Study," *Technology Science* (2017); B. Malin and L. Sweeney, "Re-Identification of DNA through an Automated Linkage Process," *Proceedings: AMIA Symposium* (2001): 423–427.

57. Lauren Sarkesian and Spandana Singh, "How Data Brokers and Phone Apps Are Helping Police Surveil Citizens Without Warrants," *Issues in Science and Technology,*

January 6, 2021, https://issues.org/data-brokers-police-surveillance/; "Data Broker Helps Police See Everywhere You've Been with the Click of a Mouse: EFF Investigation" (press release), *Electronic Frontier Foundation*, September 1, 2022, https://www.eff.org/press/releases/data-broker-helps-police-see-everywhere-youve-been-click-mouse-eff-investigation.

58. E.g., Zuboff, *The Age of Surveillance Capitalism: The Fight for a Human Future at the New Frontier of Power* (New York: Public Affairs, 2018); Eric A. Posner and E. Glen Weyl, "Data as Labor: Valuing Contributions to the Digital Economy," in *Radical Markets: Uprooting Capitalism and Democracy for a Just Society* (Princeton, NJ: Princeton University Press, 2018), 205–249.

59. Carissa Véliz, *Privacy Is Power: Why and How You Should Take Back Control of Your Data* (New York: Melville House, 2021), 87.

60. Véliz, *Privacy Is Power,* 97.

61. Of course, much of this argument also appears in Weber, and is reflected in the common experience of any middle schooler.

62. Michel Foucault, *Discipline and Punish*, 191. Gilles Deleuze's concerns about "the control" society, and Bernard Harcourt's arguments about "the expository society," as attempts to explain the forms that knowledge-power took after the end of the disciplinary society, also fit into this category.

63. Igo, *Known Citizen.*

64. Michel Foucault, *The History of Sexuality*, vol. 1, *An Introduction* (New York: Vintage, 1990), 141–242.

65. Foucault, *Discipline and Punish*, 190.

66. The nineteenth-century penny press initiated a huge increase in both the amount and the type of information about human life that made it into print. I belong to this tradition, if only by intuition, when I reach for the language of privacy to describe what my neighbor does when she documents my coming and going because of the relative fixity of writing (or the photograph) compared to ordinary living memory. Thus it is a relief for her to destroy her diaries but not for her to explain that she never recorded me doing anything out of the ordinary.

67. Although I think the view I will give is present in Foucault, I hasten to say that I am not interpreting Foucault but, as he would have wished, reading him somewhat against the grain.

68. Foucault, *History of Sexuality,* vol. 1, 98; Michel Foucault, *The History of Sexuality Volume 2: The Uses of Pleasure* (New York: Vintage, 1988), 8.

69. I noted in chapters 1 and 2 how widespread some version of this idea is in ethical and political accounts of selfhood in the work of theorists who would seem to share little in terms of foundational commitments about the metaphysics of the self—for instance from Charles Taylor's radical reevaluations to Judith Butler's concept of parodic performativity. Judith Butler, *Gender Trouble: Feminism and the Subversion of Identity* (New York: Routledge, 1999).

70. Édouard Glissant and Betsy Wing, *Poetics of Relation* (Ann Arbor: University of Michigan Press, 1997), 191.

71. Zach Blas, "Opacities: An Introduction," *Camera Obscura: Feminism, Culture, and Media Studies* 92 (2016): 149.

72. Foucault often described resistance as the "odd term of power," an essential element of it and not a transcendental opposition; Glissant describes non-dialectical resistance as "consumed with the sullen jabber of childish refusal, convulsive and powerless." Glissant and Wing, *Poetics of Relation*, 191.

73. Diary entry of October 15, 1921, in Franz Kafka, *The Diaries, 1910–1923* (New York: Schocken, 1988), 392.

74. This comes from Cioran's preface to the forty-fifth anniversary reissue of *Sur les cies du désespoir* (*On the Heights of Despair*) quoted in Marie Darrieussecq, *Sleepless* (London: Fitzcarraldo, 2023), 31. Many other examples of the link between the two oblivions can be found in Darrieussecq's profound meditation on sleeplessness.

75. Jorge Luis Borges, *Borges at Eighty: Conversations*, ed. Willis Barnstone (New York: New Directions, 2013).

76. E. M. Cioran, *On the Heights of Despair*, trans. Ilinca Zarifopol-Johnston (Chicago: University of Chicago Press, 1996).

77. Cioran, *On the Heights*, 83.

78. Alphonso Lingis, "Translator's Introduction," in *Existence and Existents* by Emmanuel Levinas (The Hague: Martinus Nijhoff: 1978), 10.

79. Nietzsche, *Unfashionable Observations*, 89.

80. Willis Barnstone, "Thirteen Questions: A Dialogue with Jorge Luis Borges," *Chicago Review* (Winter 1980): 16.

5. Privacy and the Production of Human Depth

1. Vladimir Nabokov, *Speak, Memory: An Autobiography Revisited* (New York: Vintage, 2012), 108.

2. Jenny Diski, "Trying to Stay Awake," *London Review of Books*, July 2008.

3. Wallace Stevens, "Two Letters," in *Stevens: Collected Poetry and Prose*, ed. Frank Kermode and Joan Richardson (New York: Library of America, 1997), 468–469.

4. Hannah Arendt, *Men in Dark Times* (San Diego, CA: Harcourt, Brace, 1995), 201.

5. Hannah Arendt, *The Human Condition*, 2nd ed. (Chicago: University of Chicago Press, 1998), 6.

6. Arendt, *Human Condition*, 5.

7. Arendt, *Human Condition*, 71. See also p. 57: "The reality of the public realm relies on the simultaneous presence of innumerable perspectives and aspects in which the common world presents itself and for which no common measurement or denominator can ever be devised."

8. Arendt, *Human Condition*, 57.

9. Arendt, *Human Condition*, 95.

10. Arendt, *Human Condition*, 198–199.

11. Arendt, *Human Condition*, 41.

12. Arendt, *Human Condition*, 38. The metaphorical use of ancient privacy practices is again revealed by the fact that, as Arendt well knew, "privacy" was not a word they would have used.

13. Arendt, *Human Condition*, 71.

14. Arendt, *Human Condition*, 71.

15. E.g., Shiraz Dossa, *The Public Realm and the Public Self: The Political Theory of Hannah Arendt* (Waterloo, Canada: Wilfrid Laurier University Press, 1989).

16. Arendt, *Human Condition*, 51.

17. Hannah Arendt, "The Crisis in Education," in *Between Past and Future: Eight Exercises in Political Thought* (New York: Penguin, 2006), 183.

18. Seyla Benhabib, *The Reluctant Modernism of Hannah Arendt* (Thousand Oaks, CA: Sage, 1996), 212.

19. Arendt, *Human Condition*, 6.

20. Arendt, *Human Condition*, 51–52. She also thought there were reasons, based in the value of the public realm to human and political life, in favor of confining love and other strong emotions to the private realm. Love, she thought, sought to create union out of plurality, which made it inimical to the valuable pluralism of the public sphere but also dangerous and fickle as a principle of politics which, often as not, resulted in the world alienation she sought to correct (for example, Christian love and the fascist idea of the ethno-nationalist family).

21. I borrow the phrase "strong misreading" from Harold Bloom, *The Anxiety of Influence: A Theory of Poetry*, 2nd ed. (New York: Oxford University Press, 1996).

22. Hannah Arendt, *The Life of the Mind* (San Diego, CA: Harcourt, 1981), 31.

23. Arendt, *Life of the Mind*, 32.

24. Hannah Arendt, *Life of the Mind*, 31. She makes this point in other works where she does not resort to the conceptual division between soul and mind. For instance, in *The Human Condition*, Arendt writes: "What claims our attention is the *veritable gulf* that separates all bodily sensations, pleasure or pain, desires and satisfactions—which are so 'private' that they cannot be adequately voiced, much less represented in the outside world, and therefore are altogether incapable of being reified—from mental images which lend themselves so easily and naturally to reification that we neither conceive of making a bed without first having some image, some 'idea' of a bed before our inner eye, nor can we imagine a bed without having recourse to some visual experience of a real thing" (141).

25. Arendt, *Human Condition*, 50–51. Here Arendt recalls Wittgenstein's *Philosophical Investigations*, published just five years earlier, in 1953. For example, "[Pain] is not Something, but not a Nothing either. The conclusion was only that a Nothing would render the same service as a Something about which nothing could be said." Ludwig Wittgenstein, *Philosophical Investigations*, trans. P. M. S. Hacker and Joachim Schulte, rev. 4th ed. (Chichester, UK: Wiley-Blackwell, 2009), 108–9. Also Terry Eagleton: Pain "is part of the body's obdurate resistance to intelligibility, its blind, obtuse persistence in its own being." *The Gatekeeper* (London: Penguin, 2001), 112–113.

26. Note that actually being in public isn't necessary for the destruction of the nonlinguistic experiences "of the soul," only articulating them into language and information.

This lends support to the arguments of chapter 3 about what is lost when the condition of privacy approaches the condition of hiding.

27. Arendt, *Human Condition*, 62.

28. Arendt, *Human Condition*, 62–63.

29. William Shakespeare, "King Lear," in *The Riverside Shakespeare: The Complete Works*, 2nd ed. (Boston: Houghton Mifflin, 1997), Act I, Scene 1.

30. Julie Inness, *Privacy, Intimacy, and Isolation* (New York: Oxford University Press, 1996); Charles Fried, *An Anatomy of Values: Problems of Personal and Social Choice* (Cambridge, MA: Harvard University Press, 1970); James Rachels, "Why Privacy Is Important," *Philosophy and Public Affairs* 4, no. 4 (1975): 323; Robert S. Gerstein, "Intimacy and Privacy," *Ethics* 89, no. 1 (1978): 76–81; Jean L. Cohen, *Regulating Intimacy: A New Legal Paradigm* (Princeton, NJ: Princeton University Press, 2004).

31. E.g., Gerstein, "Intimacy and Privacy."

32. This conception of the right to privacy comes Thomas Cooley's 1878 *A Treatise on the Law of Torts* and is cited, among many other places, in "The Right to Privacy."

33. Arendt, *Human Condition*, 71.

34. Arendt, *Human Condition*, 62–63.

35. Arendt, *Human Condition*, 63.

36. This is an idea I first explored in a short article: Lowry Pressly, "Requiem for the Stranger," *Political Theory* 51, no. 1 (2023): 224–233. The distinction between round and flat characters is now a cliché of fiction-writing manuals and workshops, but it originates, as far as I can tell, with E. M. Forster, *Aspects of the Novel* (San Diego, CA: Harcourt Brace, 1985). See, also, Marta Figlerowicz, *Flat Protagonists: A Theory of Novel Character* (New York: Oxford University Press, 2016). It is worth noting, too, that the roundness or depth of a character is of special importance in the genre of *realism*; this is because although it is an illusion, it is one meant to mimic or reproduce an aspect of life as it really is off the page.

37. Forster, *Aspects of the Novel*, 78.

38. This person is unfree in part because they have "contextual integrity," which is what many think privacy is for. See, e.g., Helen Fay Nissenbaum, *Privacy in Context: Technology, Policy, and the Integrity of Social Life* (Stanford, CA: Stanford Law Books, 2010). There are other good arguments against contextual integrity, several of which I make in this book (in addition to Foucauldian-type questions of discipline and domination it raises), but there is something profound in the example here, I think, about how a life defined by always being in sync with the expectations of what is appropriate to disclose in a given situation (expectations that are imposed upon the individual) would be a life appropriately described as flat or shallow. "Live a little!" we want to say to such a person, for their life seems to lack the quality of surprise and liveliness that comes from the darker ground of personal oblivion. Such a person's conformity with public expectations seems to cast doubt on whether he has such a darker ground at all or whether, I want to say, it is working.

39. Arendt, *Human Condition*, 63.

40. Luke 8:17 (New International Version).

41. E.g., Michel Foucault and Frédéric Gros, *The Hermeneutics of the Subject: Lectures at the Collège de France, 1981–1982*, 1st ed., Lectures at the Collège de France (New York: Picador, 2006).

42. Alain Corbin, "Backstage," in *A History of Private Life*, vol. 4 (Cambridge, MA: Belknap Press of Harvard University Press, 1990), 460.

43. See, e.g., Wolfgang Schivelbusch, *The Railway Journey: The Industrialization of Time and Space in the Nineteenth Century*, with a new preface (Oakland, CA: University of California Press, 2014).

44. Deborah Lupton, *The Quantified Self: A Sociology of Self-Tracking* (Cambridge, UK: Polity, 2016).

45. Deborah Lupton, "You Are Your Data: Self-Tracking Practices and Concepts of Data," in *Lifelogging: Digital Self-Tracking and Lifelogging—between Disruptive Technology and Cultural Transformation*, ed. Stefan Selke (Wiesbaden, Germany: Springer Fachmedien Wiesbaden, 2016), 61–79.

46. Arendt, *Human Condition*, 64.

47. Arendt, *Human Condition*, 63.

48. Marcel Proust, *In Search of Lost Time, Volume I: Swann's Way*, trans. C. K. Scott-Moncrieff, Terence Kilmartin, and D. J. Enright (New York: Modern Library, 2003), 312. It is worth noting that the "personal goods" above are not strictly individual, but depend on common structures of the material and ideological landscape—like, say, the benefit one gets grazing one's sheep on the town common. Because they are grounded in the social and political, they are therefore properly the subject of that sort of attention, critique, and support.

49. Joshua Rothman, "Virginia Woolf's Idea of Privacy," *New Yorker*, July 9, 2014, https://www.newyorker.com/books/joshua-rothman/virginia-woolfs-idea-of-privacy.

50. Arendt, *Life of the Mind*, 21.

51. Richard Sennett presents a similar view of the meaningfulness of theatrical expression in his own lament for the loss of a public realm more conducive to human flourishing in *The Fall of Public Man* (New York: Norton, 1996).

52. Arendt, *Human Condition*, 51.

53. Appearance as such must also surely derive some of its vitality, its sheer *thereness* and the hold it has over us, from its appearing against a background of oblivion (as opposed to a pseudo "darker ground" that is either invisible or opaque, or behind which are thought to exist not oblivion but rather more appearances that are merely hidden). This is, I think, something in the area of what John Berger thought about the almost religious luminosity of even secular painting: "Painting is, first, an affirmation of the visible which surrounds us and which continually appears and disappears. Without the disappearing, there would perhaps be no impulse to paint, for then the visible itself would possess the surety (the permanence) which painting strives to find. More directly than any other art, painting is an affirmation of the existent, of the physical world into which mankind has been thrown." John Berger, "Steps towards a Small Theory of the Visible (for Yves)," in *The Shape of a Pocket* (New York: Knopf, 2009), 14.

54. *Pace* Plato.

55. See, e.g., Karen Jones, "The Politics of Intellectual Self-Trust," *Social Epistemology* 26, no. 2 (2012): 237–251; Richard Foley, *Intellectual Trust in Oneself and Others* (Cambridge: Cambridge University Press, 2001).

56. Ralph Waldo Emerson, "The American Scholar," in *The Essential Writings of Ralph Waldo Emerson* (New York: Modern Library, 2000), 55.

57. Charles Taylor, "What Is Human Agency?," in *Human Agency and Language*, Philosophical Papers 1 (Cambridge: Cambridge University Press, 1985), 41–42.

58. The phrase is Arendt's, from Hannah Arendt, *On Revolution* (New York: Penguin, 2006), 194.

59. Ralph Waldo Emerson, "Self-Reliance," in *The Essential Writings of Ralph Waldo Emerson* (New York: Modern Library, 2000), 138.

60. Emerson, "Self-Reliance," 140.

61. Emerson, "Self-Reliance," 141.

62. Francis Fukuyama, *Trust: The Social Virtues and the Creation of Prosperity* (New York: Free Press, 1995).

63. Joseph Kupfer, "Privacy, Autonomy, and Self-Concept," *American Philosophical Quarterly* 24, no. 1 (1987): 85.

64. Kupfer, "Privacy, Autonomy, and Self-Concept," 85. Indeed, a judgment that K has no right to be forgotten would be tantamount to judging him untrustworthy. Since his conviction is not of any real historical import, at least not of the sort that requires his real name be attached to it, the public interest that weighed against his right to be forgotten was one in public safety.

65. This suggests another hypothesis for why the massive increase in connectivity in the digital age left individuals feeling more isolated and alienated than ever. The technology of mobile connectivity is also a technology of near-constant mutual surveillance: we keep up with others and know they are keeping up with us, and we know that our movements and activity are being tracked by myriad corporations to a variety of ends. If diminished opportunities for not being tracked leads to a diminished sense of trustworthiness—and if a sense of trustworthiness is important for both an individual's self-worth and her positive connection to society—then we might expect an increase in tracking to contribute to an increase in social isolation.

66. See, e.g., the discussion of moral, civic, and psychological development of children denied certain opportunities to act well (and to be trusted) in Annette Baier, "Demoralization, Trust, and the Virtues," in *Reflections on How We Live* (Oxford: Oxford University Press, 2010).

67. Kupfer, "Privacy, Autonomy, and Self-Concept," 85.

68. Michel Foucault, *Discipline and Punish: The Birth of the Prison* (New York: Vintage, 1995); Michel Foucault, *The History of Sexuality*, vol. 1, *An Introduction* (New York: Vintage, 1990); Max Weber, *The Protestant Ethic and the "Spirit" of Capitalism and Other Writings* (New York: Penguin, 2002); Erving Goffman, *Asylums: Essays on the Social Situation of Mental Patients and Other Inmates* (New York: Anchor, 1990), 1.

69. Thomas Nagel, "Concealment and Exposure," in *Concealment and Exposure: And Other Essays* (Oxford: Oxford University Press, 2002), 15.

Acknowledgments

I cannot shake the conviction that the words and ideas in this book cannot be said to be mine except in the most trivial sense—as a stone might say "mine" about the river that passes around and over it, altering its form over time and filling it with holes and unexpected crevices through which the water runs on its way to unreachable places in the distance—notwithstanding what might appear on the copyright page. So, although I am responsible for this book, I wish to acknowledge the many sources that have brought it to me, in bits and pieces, in friendship and wanton generosity, so that it could pass through me and into your hands, reader (whom I also wish to thank for your time and attention). This was a difficult book to write, and I could not have even begun without an enormous amount of aid and support.

This book would have been impossible without the constant guidance and inspiration of Michael Sandel. On every page it bears the mark of his persistent and generous questioning and encouragement. Likewise, I owe enormous gratitude to Nancy Rosenblum, Eric Beerbohm, and Danielle Allen, with whom I have discussed these ideas for over a decade now and who have influenced the substance and style of this book in innumerable, fundamental ways. Thanks also to Bernardo Zacka, who helped me many times over the years to see these arguments more broadly and deeply, and who always seemed to understand this book better and more eloquently than I did.

This book is the product of years of animated debate and enriching conversations with many more friends and readers, of whom I can only mention a fraction. My lifelong gratitude to Jacob Abolafia, Rachel Achs, Adriana Alfaro Altamirano, Avishay Ben Sasson-Gordis, Matthew Boyle, James Brandt, Carmen Dege, Sandy Diehl, Leah Downey, David Estlund, Tweedy Flanigan,

Jennifer Forestal, Katrina Forrester, Jonathan Gould, John Harpham, Bonnie Honig, Sergio Imparato, Tae-Yeoun Keum, Ian Malcolm, Richard Moran, Eric Nelson, Zeynep Pamuk, Charles Petersen, Michael Rosen, Adam Sandel, Josh Simons, David Skarbek, Lucas Swaine, and many others.

I am grateful for the support and lively exchange of ideas I have received from the Edmond and Lily Safra Center for Ethics at Harvard; the Center for Philosophy, Politics, and Economics at Brown; the Committee on Degrees in Social Studies at Harvard; and the Chronophages. Thanks to my students, too, for constantly challenging me with new ideas and fresh perspectives. And to Dr. Jorge Arredondo in Mexico City and the surgeons and physical therapists at University Orthopedics in East Providence, RI, for taking care of me when I broke both hands while writing this book (during its composition, that is, not as a result of it).

My deepest gratitude to everyone at Harvard University Press, especially Joseph Pomp, whose combination of vision, rigor, and style, and whose belief in this book and demand that it live up to the standard to which it aspired, are evident on every page. It is a joy to collaborate with such a thoughtful and ambitious editor. My thanks also to Jillian Quigley, Stephanie Vyce, and Susan Virtanen, whose hard work and careful attention were crucial in getting this book to print, and to the two anonymous reviewers for their insightful readings and thoughtful suggestions.

Thanks also to Arne Svenson for so generously allowing his images to be used in the book. Chapter 4 expands on work presented in "The Right to Be Forgotten and the Value of an Open Future," *Ethics* 135, no. 1 (2024). Chapter 5 touches on ideas first presented in "Requiem for the Stranger," *Political Theory* 51, no. 1 (2023), 224–233.

Finally, I must thank my parents, who made many sacrifices so that I could get an education and write books. And Regina, without whom I would not bother to write them.

Index

accountability: agency and, 19, 71–83, 86, 118, 174, 196n20; human depth and, 174; invasions and, 19, 71, 77–83, 86, 118, 174, 196n20; memory and, 118; moral and, 73, 76, 80–83

Acts of Oblivion, 16–17

addiction, 109, 114

agency: accountability and, 19, 71–83, 86, 118, 174, 196n20; Arendt and, 197n33; attention and, 69, 71, 74, 83, 86; autonomy and, 63, 74, 77, 81, 84; boundaries and, 86; cultural issues and, 81; emotion and, 83; ethics and, 74; Foucault and, 194n13, 195n15, 196n21; harm and, 58–69, 72, 81, 193nn3–4, 194nn8–9, 195n16; human depth and, 149–150, 162, 169–175; identity and, 69, 74, 196n20; individuality and, 77; Internet and, 70; intimacy and, 82–87, 197n16, 198n41; invasion and, 58, 71, 194n8; memory and, 123, 133, 137–138, 143, 145, 147, 205n23; moral issues and, 73–77, 80–81; objectivity and, 193n6; oblivion and, 15, 19, 21; opacity and, 80, 82; personality and, 73, 77, 195n14, 197n27; philosophers and, 71, 74, 195n17, 196n21, 197n32, 198n42; photography and, 39–41, 44, 53; politics and, 15, 74, 81–82; psychology and, 63, 74, 77, 79,

86; publicity and, 72–75; right to privacy and, 60–62, 84; secrets and, 58–61, 71, 80, 82, 85, 194n8; self-knowledge and, 74–79, 87, 195n17; self-relation and, 15, 19, 80, 197n33; solitude and, 82–83; subjectivity and, 196n20; surveillance and, 58–60, 83; technology and, 63, 66, 85; unaccountability and, 71–72, 76–87, 195n16; violations and, 58–72

Alcoholics Anonymous, 12–13

Alcott, Amos Bronson, 192n61

alienation, 20, 83, 93, 113–117, 151, 155, 211n20, 214n65

anonymity: census data and, 192n61; memory and, 128–131, 134, 138–139; photography and, 28, 56; privacy and, 12–13, 15

anxiety, 10–11, 28, 31, 90, 110

Arendt, Hannah: "The Crisis in Education," 115; darkness and, 153, 159–160, 165, 167; hiding and, 93, 102, 115–116; *The Human Condition*, 116, 150–151, 155, 158, 211n20, 211n24; human depth and, 21, 150–163, 165–168, 173, 211n12, 211nn24–25; identity and, 166–167, 210n10; isolation and, 93, 102, 116; Kupfer and, 173–174; *The Life of the Mind*, 155–156, 211n24; love and, 155–158, 161; objectivity and, 115, 152, 155,

Arendt, Hannah (*continued*)
    163, 167, 173; ontology of privacy and,
    151–154, 192n63; personal depth and,
    160–164; public goods and, 166–167;
    publicity and, 151–152; self-relation and,
    155–157; self-trust and, 173–174; totalitari-
    anism and, 93, 116, 151
Aristotle, 195n15
attention: agency and, 69, 71, 74, 83, 86;
    hiding and, 88–93, 99–109, 114; human
    depth and, 148, 153–154, 211n24, 213n48;
    information age and, 4, 16; memory and,
    120, 124, 129, 135, 138–142, 207n37,
    207n43; photography and, 24–25
autonomy: agency and, 63, 74, 77, 81, 84;
    hiding and, 106, 112; human depth and,
    162–163, 171; memory and, 125, 138, 141;
    oblivion and, 6; photography and, 31;
    voyeurism and, 63, 74, 77, 81, 84

Barthes, Roland, 51–53, 104, 137
Benhabib, Seyla, 154–155
Berger, John, 213n53
biographical dimension, 66–67, 70, 72, 76,
    194n11
biometrics, 7, 10
blackmail, 14, 139
Blackstone, William, 189n25
Blake, William, 155
Blas, Zach, 143–144
bodily integrity, 6
Bok, Sissela, 184n19
Borges, Jorge Luis: courage and, 20, 147;
    "Funes, El Memorioso", 121–129, 133–136,
    142–147, 150, 159–165, 172, 203n4, 204n12,
    204n14, 207n43; human depth and, 150,
    159–165, 172; imagination and, 20–21, 147;
    memory and, 20–21, 121–128, 142,
    145–147, 204n12, 204n14, 205n17, 207n43;
    sleep and, 124, 150
boundaries: agency and, 86; Eleusinian
    mysteries and, 162–163; hiding and, 112,
    115; human depth and, 163; photography

and, 25–26, 29, 41, 47, 54–55, 191n59;
    self-knowledge and, 168; social norms
    and, 13
Brandeis, Louis, 27, 34–35, 38–41, 46–53, 90,
    154, 187n18, 189n25, 190n34, 190n42,
    191n53, 192n61, 192n73, 194n9; newspa-
    pers and, 90; on property, 48–49;
    photography and, 27, 34–35, 38–41, 46–53,
    187n18, 189n25, 190n34, 190n42, 191n53,
    192n61, 192n73; publicity and, 90, 154;
    "The Right to Privacy" and, *see* "Right to
    Privacy, The"; Schiller and, 46, 187n18;
    spirituality and, 48, 194n9
branding, 2, 134–135, 208n44
Browne, Simone, 82
Butler, Judith, 196n20, 209n69

Caché (film), 174
Camera Lucida (Barthes), 51–52
capitalism, 5–7, 139, 181n6, 198n36
Cavell, Stanley, 101
CCTV, 19
census data, 164, 192n61
Cioran, Emil, 145
clues, 16
Commentaries (Blackstone), 189n25
confidentiality, 20, 184n26, 185n10,
    191n59
connectivity: alienation and, 114–117; hiding
    and, 89, 92–93, 103, 106–115; human
    depth and, 214n65; increase in, 214n65;
    information age and, 10–11, 89, 92–93,
    103, 106–117, 131, 214n65; memory and,
    131; mobile, 10–11, 92, 103, 114, 131, 214n65;
    social media and, 11, 92, 103; tethered,
    100–104, 106–109; Turkle on, 89
consent: Foos and, 72, 195n16; invasion and,
    5, 11, 25, 45, 56, 72, 91, 105, 108, 195n16;
    moral issues and, 5, 11, 195n16; photog-
    raphy and, 45–46, 56
conversation: audience and, 110; chat and,
    103–104; connectivity and, 89, 202n45;
    face-to-face, 111; FOMO and, 108;

overhearing, 88, 186n12; photography and, 46; privacy of home and, 187n16; scrolling and, 202n45

Cooley, Thomas, 212n32

Creeley, Robert, 196n20

daguerreotypes, 44, 191n53

darkness: Arendt and, 146, 153, 158, 159–160, 165, 167; Berger on, 213n53; hiding and, 89, 107–110, 114; human depth and, 153–160, 165–172, 212n38, 213n53; information age and, 20; insomnia and, 145; memory and, 145–146; oblivion and, 3, 9, 16, 146, 159; self-trust and, 172–173; Shakespeare on, 16; Stevens on, 9; surveillance and, 172–173

Darrieussecq, Marie, 210n74

Darwin, Charles, 32–33

databases, 44; archives and, 30, 119–120, 128–132, 137, 142–143, 164, 169, 190n38; documentation and, 17, 21, 119, 136–144, 175, 183n16, 194n13

datafication, 21, 119

data subjects, 1, 181n1

deep fakes, 36

Deleuze, Gilles, 209n62

Delphi, 2, 125, 147

democracy, 2, 114, 139

depth, 160 (see also human depth)

digital age: ethics and, 8, 10, 164, 177–178; hiding in, 89, 93, 95, 110–114; human depth and, 164; memory and, 128; oblivion and, 1, 8–10; photography and, 36, 44

Diski, Jenny, 148–149

documentation: datafication and, 21, 119; information gathering and, 17; memory and, 17, 21, 119, 136–144, 175, 183n16, 194n13; rampant use of, 141–142, 194n13

dreams, 122, 148–150, 155, 181n3, 188n24

Duchenne, Guillaume, 32

Eastman, George, 45

Ellison, Ralph, 12, 183n18

Emerson, Ralph Waldo, 32, 46, 52, 169, 171, 192n61

emotion: agency and, 83; anxiety, 10–11, 28, 31, 90, 110; human depth and, 155–156, 165, 211n20; photography and, 33, 38, 46, 51; tethering and, 93

ethics: agency and, 74; digital age and, 8, 10, 164, 177–178; fixity and, 137–144; Foucault and, 142; hiding and, 92, 105, 113; human depth and, 160, 163–164, 167, 174–175; information age and, 2, 8–11, 13, 17, 122, 143, 164, 178; Internet and, 8, 10, 164; memory and, 122, 137, 142–144; oblivion and, 2–13, 17; photography and, 30–31, 51, 56; politics and, 2–13, 51, 56, 92, 122, 143–144, 163–164, 174, 177–178, 181n6, 182n7; self-knowledge and, 2; technology and, 6, 8, 10–11, 142–144, 160, 164, 177–178

exhibitionism, 25, 84–86, 91–92

Facebook, 7–8

facial expression, 32–33, 45–47

facial recognition, 11, 18–19

Farahany, Nita, 50–51

Federal Constitutional Court of Germany, 187n18

feminist critiques of privacy, 81–82, 153, 197n35, 198n36

fingerprints, 137

First Amendment, 25

Fitbit, 120

FOMO, 108

Foos, Gerald: consent and, 72, 195n16; harm and, 19, 58–64, 67–69, 72, 193n4; Manor House Motel and, 58–59; moral issues and, 58–63, 69, 73, 76, 195n16; rationalization of, 58–59; spying of, 19, 58–64, 67, 70, 79, 194n8, 195n16; voyeurism of, 19, 58–73, 76–80, 95, 193n4, 195n16, 197n34

forgetting: archives and, 30, 119–120, 128–132, 137, 142–143, 164, 169, 190n38; Borges on, 121–129, 133–136, 142–147, 150, 159–165, 172, 203n4, 204n12, 204n14,

forgetting (*continued*)
207n43; criminal branding and, 130–137; documentation and, 17, 21, 119, 136–144, 175, 183n16, 194n13; fixity and, 137–144; Foucault and, 134, 136; hiding and, 112; human depth and, 149, 162, 173; insomnia and, 64, 123, 126, 145–148; memory and, 118–124, 127–129, 132, 134, 146, 204n9, 206n36; oblivion and, 13, 15, 21–22, 184n19; photography and, 39; story of K and, 130–137; unendurable precision of being and, 119–127

Forster, E. M., 161

Foster v. Svenson, 184n23, 185n1, 188n24

Foucault, Michel: agency and, 196n21; autobiographical writing and, 164; *Discipline and Punish*, 136, 140, 194n13; ethics and, 142; documentation and, 140, 175, 194n13; forgetting and, 134, 136; Glissant and, 144; identity and, 53, 134, 143, 147; individuality and, 194n13; memory and, 134, 136, 209n62, 209nn67–68, 210n72; nominalism of, 53; odd term of power, 210n72; panopticism and, 140, 175; personality and, 134, 136; personal knowledge and, 141–142; subjectivity and, 53

Fox Talbot, William Henry, 190n34

Frankfurt, Harry, 78, 197n25

Freitas, Donna, 109–110

Freud, Sigmund, 99, 120, 200n22

"Funes, El Memorioso" (Borges): forgetting and, 121–129, 133–136, 142–147, 150, 159–165, 172, 203n4, 204n12, 204n14, 207n43; human depth and, 150, 159–165, 172; sleep and, 124, 150; tragedy of, 121–129

general data protection regulation (GDPR), 183n12

genocide, 128

Glissant, Édouard, 82, 143–144, 209n72

God, 164

Goffman, Erving, 175, 197n33

Google, 129, 205n23

GPS tracking, 19, 172

Greenwald, Glenn, 58

Halpern, Sue, 50

Haneke, Michael, 174

Harcourt, Bernard, 4–5, 89–90, 112, 138, 175, 209n62

harm: agency and, 58–69, 72, 81, 193nn3–4, 194nn8–9, 195n16; confidentiality and, 184n26; definition of, 193n3; hiding and, 90, 98, 108; human depth and, 153, 156, 158, 169; information access and, 4, 183n12; legal issues and, 4, 19, 58–64, 67–69, 72, 193nn3–4; media and, 38, 90, 108; memory and, 119, 131–140; newspapers and, 90; photography and, 29, 38–40, 48–49, 53; voyeurism and, 19, 58–69, 67–69, 72, 81, 193n4; well-being and, 61–64

Hawthorne, Nathaniel, 29

health apps, 120

Heisenberg, Werner, 9

Herostratus, 132

Herzog, Werner, 3

hiding: abjection of, 98–100; alienation and, 114–117; Arendt and, 93, 102, 115–116; attention and, 88–93, 99–109, 114; autonomy and, 106, 112; blurred lines in, 111–114; boundaries and, 112, 115; concept of, 92; connectivity and, 89, 92–93, 103, 106–115; cultural issues and, 92, 105, 107, 114, 200n18, 200n23; darkness and, 89, 107–110, 114; digital age and, 89, 93, 95, 110–114; ethics and, 92, 105, 113; forgetting and, 112; Franks and, 94–98, 101, 107, 199n15; fugitives and, 94; harm and, 90, 98, 108; identity and, 97, 101–105, 199n15; individuality and, 101; Internet and, 95, 106–107, 110–111, 114; intimacy and, 109; invasion and, 90, 105, 107; isolation and, 20, 93, 101–102, 106, 108, 114–117, 202n45; legal issues and, 200n23; media and,

89–90, 92, 108–109, 199n10, 202n41; moral issues and, 91–99, 110–113, 199n10, 199n15, 200n23, 202n50; negative connotation of, 200n23; newspapers and, 89–90, 103, 110–111, 199n10; objectivity and, 97, 115; peeping and, 107; philosophers and, 95, 112, 200nn17–18; politics and, 89, 92, 97, 116; psychology and, 93, 99–104, 109, 112; publicity and, 90–91, 105, 111, 114–116, 199n10; radio and, 91, 93, 103, 111, 199n10; relation to privacy, 94–98; right to privacy and, 89; secrets and, 94–100, 104–105, 110, 199n15, 200n22; self-knowledge and, 102; self-relation and, 104–106; smartphones and, 88–89, 93, 95, 105–111, 114, 202n45; social media and, 89, 92, 103, 107–110, 202n41; solitude and, 93, 102–106, 109, 111, 116; subjectivity and, 89, 107, 112, 114; surveillance and, 91–92, 104; technology and, 89–90, 103, 107–110, 113–114, 199n15; tethering and, 100–110, 113, 116–117; violations and, 97, 104–105, 111

Holmes, Oliver Wendell, 35, 44, 119, 132, 189n31, 190n34

hubris, 122, 149, 205n17

human depth: accountability and, 174; agency and, 149–150, 162, 169–175; Arendt and, 150–163, 165–168, 173, 211n12, 211nn24–25; attention and, 148, 153–154, 211n24, 213n48; autonomy and, 162–163, 171; Borges on, 150, 159–165, 172; boundaries and, 163; connectivity and, 214n65; cultural issues and, 164, 173; darkness and, 153–160, 165–172, 212n38, 213n53; digital age and, 164; dreams and, 122, 148–150, 155, 181n3, 188n24; emotion and, 155–156, 165, 211n20; ethics and, 160, 163–164, 167, 174–175; forgetting and, 149, 162, 173; harm and, 153, 156, 158, 169; identity and, 148–152, 166–167, 171; ideology and, 172, 213n48; individuality and, 153; Internet and, 164, 172; intimacy and, 157; isolation and, 214n65; Kupfer and, 173–174; media and, 164; moral issues and, 154, 157, 164, 171–174, 214n66; objectivity and, 148, 152, 155, 163, 167, 173; opacity and, 153, 174; personal depth and, 160–164; personality and, 155, 161; philosophers and, 153, 160, 163; politics and, 150–157, 163–167, 174, 211n20, 213n48; psychology and, 148, 172, 214n66; publicity and, 151–159, 166, 175; right to be forgotten and, 165, 171, 205n23, 205n27, 206n36, 207n43, 214n64; right to privacy and, 158; secrets and, 162–163, 166, 170, 172, 175; self-knowledge and, 149, 157, 160, 162, 165, 168–171; self-relation and, 155–157, 170; self-trust and, 9, 102, 167–172, 175; shallowness and, 2, 22, 147, 150, 158–161, 164, 169, 176, 212n38; sleep and, 148–150, 172–173; social media and, 164; solitude and, 162, 166; subjectivity and, 159, 164; surveillance and, 164, 172, 214n65; technology and, 149, 151, 160–161, 164, 172–173; trustworthiness and, 167–176; unaccountability and, 162, 174; violations and, 158

human rights, 5, 13, 69, 164

Hurley, Andrew, 204n14

identity: agency and, 69, 74, 196n20; Arendt and, 166–167, 210n10; cultural issues and, 2, 4, 105, 127; Foucault and, 143; hiding and, 97, 101–105, 199n15; human depth and, 148–152, 166–167, 171; identity and, 2, 4, 105, 127; memory and, 127, 133–134, 141–147; personal, 19, 69, 74, 102, 133–134, 143–150; photography and, 36, 53; sleep and, 148; unaccountability and, 69, 74

ideology: blinded by, 12; capitalism, 5–7, 139, 181n6, 198n36; democracy, 139; human depth and, 172, 213n48; information age and, 3–13, 181n6, 183n16; memory and, 136, 144; neoliberalism, 4–5, 91, 138;

ideology (*continued*)
　　photography and, 51, 187n18; plutocracy,
　　139–140
ignorance, 9, 134, 153, 192n71
Igo, Sarah, 137–138
individuality: agency and, 77 (*see also*
　　agency); Barthes on, 52; Foucault on,
　　194n13; hiding and, 101; human depth
　　and, 153; memory and, 131, 140, 143, 146;
　　Romanticism and, 30
information age: blinded by, 2; connectivity
　　and, 10–11, 89, 92–93, 103, 106–117, 131,
　　214n65; darkness and, 20; ethics and, 2,
　　8–11, 13, 17, 122, 143, 164, 178; ideology of,
　　3–13, 181n6, 183n16; impact of, 1–4;
　　memory and, 118–119, 131–132, 137, 139;
　　periodization of, 183n16; privacy and, 8,
　　10; solutions in, 177–178; tethering and, 20
informational privacy, 7, 14, 20, 95–97, 106,
　　183n10, 184n19
innovation, 10–11, 31, 190
insomnia: assault and, 64; Borges and, 123,
　　126, 145–148; sleep disturbances and, 64,
　　123, 126, 145–148
Internet: agency and, 70; ethics and, 8, 10,
　　164; hiding and, 95, 106–107, 110–111, 114;
　　human depth and, 164, 172; memory
　　and, 21, 119–120, 123, 128–132, 137–142,
　　190n38, 206n36; periodization of, 183n16;
　　photography and, 35; privacy features of,
　　7–8, 177; right to be forgotten, 11, 16, 21,
　　118–119, 127–139, 143, 165, 171, 205n27,
　　206n36, 207n43, 214n64; story of K and,
　　131
Internet of Things (IOT), 28, 164
intimacy: agency and, 82–87, 197n16, 198n41;
　　hiding and, 109; human depth and, 157;
　　kissing, 79, 84–87, 158; unaccountable,
　　83–87
intrusion, 25–26, 123, 139
invasion: accountability and, 19, 71, 77–83,
　　86, 118, 174, 196n20; agency and, 194n8;
　　biographical dimension and, 66–67, 70,

72, 76, 194n11; consent and, 5, 11, 25, 45,
　　56, 72, 91, 105, 108, 195n16; exhibitionism
　　and, 25, 84–86, 91–92; Foos and, 19,
　　58–73, 76–80, 95, 193n4, 195n16, 197n34;
　　hackers, 63–66; hiding and, 90, 105, 107;
　　intrusion and, 25–26, 123, 139; legal issues
　　and, 26; memory and, 138; newspapers
　　and, 89–90; photography and, 24–29,
　　33–35, 38–41, 44–50, 55–56, 189n27,
　　190n34; physical, 18; privacy and, 18–19;
　　by questions, 54–57; radio and, 91; right
　　to be forgotten, 11, 16, 21, 118–119, 127–139,
　　143, 165, 171, 205n27, 206n36, 207n43,
　　214n64; spying and, 17, 39, 58–64, 67, 70,
　　79, 194n8, 194n11, 195n16; stalking and,
　　108, 111, 120, 139; technology and, 18, 25,
　　27, 34–35, 39, 46, 50, 90; trespass and, 48;
　　Voyeur's Motel and, 19, 58, 71; well-being
　　and, 61–64
isolation, 28; Arendt and, 93, 102, 116;
　　hiding and, 20, 93, 101–102, 106, 108,
　　114–117, 202n45; human depth and,
　　214n65; memory and, 126, 133; Turkle on,
　　202n45
Ivins, William, 189n33

James, Henry, 90
James, William, 122
justice, 82, 200n23

Kafka, Franz, 145
Kant, Immanuel, 2, 187n18
Keats, John, 78
kissing, 79, 84–87, 158
knowledge by acquaintance, 198n42
Kodak, 45, 142
Kupfer, Joseph, 173–174

Levinas, Emmanuel, 146
liberty, 4, 6, 43, 82, 108, 206n32
Lingis, Alphonso, 145–146
loneliness, 20, 93, 109, 114–117, 162
Lorimer, George, 90–91

love, 2, 31, 33, 87, 93, 100, 134, 155–157, 162, 211n20
lovers, 14, 78–79, 84–85, 94, 105, 158, 161, 164, 173, 197n31
Lupton, Deborah, 1, 164
Luria, A. R., 203n4

MacKinnon, Catharine, 198n36
Manola, Marion, 41, 190n42
Marx, Karl, 6
Mayer-Schönberger, Viktor, 123
McLuhan, Marshall, 199n10
media: Bertillonian biometrics and, 10; harm and, 38, 90, 108; hiding and, 89–90, 92, 108–109, 199n10, 202n41; human depth and, 164; memory and, 119, 137, 142, 205n27; newspapers, 89–90 (*see also* newspapers); panopoly of, 108; penny press and, 209n66; photography and, 34, 189n27; radio, 11, 17, 91, 93, 103, 111, 125–126, 199n10; social, 5 (*see also* social media); subjectivity of, 89–90, 92; television, 11, 59, 91, 103, 111, 199n10; Turkle on, 89, 107–109, 202n45
memory: accountability and, 118; agency and, 123, 133, 137–138, 143, 145, 147, 205n23; anonymity and, 128–131, 134, 138–139; archives and, 30, 119–120, 128–132, 137, 142–143, 164, 169, 190n38; attention and, 120, 124, 129, 135, 138–142, 207n37, 207n43; autonomy and, 125, 138, 141; Borges and, 204n12, 204n14, 205n17, 207n43; Borges on, 20–21, 121–129, 133–136, 142–147, 150, 159–165, 172, 203n4, 204n12, 204n14, 207n43; connectivity and, 131; criminal branding and, 130–137; cultural issues and, 119–121, 127, 146; darkness and, 145–146; datafication and, 21, 119; digital age and, 128; documentation and, 17, 21, 119, 136–144, 175, 183n16, 194n13; eradication of, 127–128; ethics and, 122, 137, 142–144; fixity and, 137–144; forgetting and, 13, 118–124, 127–129, 132,

134, 146, 203n4, 204n9, 206n36; Foucault and, 134, 136, 209n62, 209nn67–68, 210n72; harm and, 119, 131–140; identity and, 127, 133–134, 141–147; ideology and, 136, 144; individuality and, 131, 140, 143, 146; information age and, 118–119, 131–132, 137, 139; Internet and, 21, 119–120, 123, 128–132, 137–142, 190n38, 206n36; invasion and, 138; isolation and, 126, 133; legal issues and, 129; media and, 119, 137, 142, 205n27; moral issues and, 119–122, 127–129, 133–138, 141, 144, 147, 207n43; newspapers and, 120, 130, 137, 142; Nietzsche and, 122, 127–128, 131, 142, 146, 204n9; objectivity and, 126, 142; opacity and, 143; personality and, 123, 134, 137, 144, 207n43; philosophers and, 121, 205n28; politics and, 119–122, 127–130, 136, 139, 143–144, 207n43; precision of being and, 119–127; psychology and, 118, 120, 122, 135; publicity and, 120; radio and, 125–126; Ribot and, 122; right to be forgotten, 11, 16, 21, 118–119, 127–139, 143, 165, 171, 205n27, 206n36, 207n43, 214n64; right to privacy, 129–130, 205n27; secrets and, 120, 144; self-knowledge and, 125, 142–146; social media and, 142; story of K and, 130–137; surveillance and, 136, 139–140; technology and, 119–120, 131–132, 135, 137, 141–144; tethering and, 118, 131, 143, 146; unaccountability and, 118, 145; violations and, 129–131, 205n28
mobile connectivity, 10–11, 92, 103, 114, 131, 214n65

Nabokov, Vladimir, 148
Nagel, Thomas, 175, 187n16
Neighbors, The (Svenson), 23–25, 56–57
neoliberalism, 4–5, 91, 138
Neruda, Pablo, 87
newspapers: Borges and, 120–121; Brandeis and, 90; harm and, 90; hiding and, 89–90, 103, 110–111, 199n10; invasion and,

newspapers (*continued*)
89–90; memory and, 120, 130, 137, 142; moral issues and, 33–34; penny press and, 209n66; photography and, 27, 29, 31, 34, 42, 189n27, 192n61; privacy complaints and, 10–11, 21, 89, 182n7; Rachels and, 84; story of K and, 130–133; Warren and, 90
Nietzsche, Friedrich, 122, 127–129, 131, 142, 146, 196nn20–21, 204n9
Nissenbaum, Helen, 182n9, 212n38
Nozick, Robert, 63, 193n6
NSA, 58

oblivion: collective regions of, 8; concept of, 3, 12–18; consent and, 45–46, 56; darkness and, 3, 9, 16; digital age and, 1, 8–10; ethics and, 1–13, 14, 17–21, 177, 182n7, 183n10; forgetting and, 13, 15, 21–22, 184n19; memory and, 118 (*see also* memory); origin of word, 118; public goods of, 21–22, 136, 151, 165–167; self-knowledge and, 2, 19, 22; sleep and, 126, 145, 148–150, 172; social production of, 3, 15, 21, 32, 53, 56–57, 126, 148–176
oblivious trust: Arendt and, 173–174; coming apart and, 171; darkness and, 172–173; human depth and, 167–176; Kupfer and, 173–174; problem-solving and, 167–170; self-trust and, 9, 102, 167–172, 175
opacity: agency and, 80, 82; concealment and, 13; Glissant and, 143–144; home and, 187n16; human depth and, 153, 174; information access and, 2; memory and, 143; secrets and, 13

panopticism, 136, 140, 175
paternalism, 194n8
Pavesich, Paolo, 41–44, 136, 206n32
Pencil of Nature, The (Fox Talbot), 190n34
personal data, 7, 140
personal depth, 160–164 (*see also* human depth)

personality: agency and, 73, 77, 195n14, 197n27; Foucault and, 134, 136; human depth and, 155, 161; as inviolate, 45–54, 187n18; memory and, 123, 134, 137, 144, 207n43; moral issues and, 30; right of, 187n18; Romanticism and, 30, 187n18; secrets and, 32, 47–48, 144; self-conception and, 187n17; self-knowledge and, 192n61 (*see also* self-knowledge); shallowness and, 2, 22, 147, 150, 158–161, 164, 169, 176, 212n38; spontaneity and, 31–32, 35, 52, 171, 187n16; stereotypes and, 31; violations and, 42–54, 187n18; Whitman and, 191n56
photography: agency and, 39–41, 44, 53; anonymity and, 28, 56; attention and, 24–25; autonomy and, 31; boundaries and, 25–26, 29, 41, 47, 54–55, 191n59; Brandeis and, 27, 34–35, 38–41, 46–53, 187n18, 189n25, 190n34, 190n42, 191n53, 192n61, 192n73; cultural issues and, 28, 34, 54, 188n24; daguerreotypes, 44, 191n53; deep fakes and, 36; digital age and, 36, 44; emotion and, 33, 38, 46, 51; ethics and, 30–31, 51, 56; First Amendment and, 25; forgetting and, 39; harm and, 29, 38–40, 48–49, 53; Holmes on, 35, 119; identity and, 36, 53; ideology and, 51, 187n18; Internet and, 35; intrusive, 25–26; invasion and, 24–29, 33–35, 38–41, 44–50, 55–56, 189n27, 190n34; Kodak and, 45, 142; legal issues and, 26–28, 34, 41, 53, 184n23, 185n1, 187n18, 188n24; Manola and, 41; media and, 189n27; moral issues and, 24–30, 33–34, 37–45, 53–54, 187n18, 188n24; newspapers and, 27, 29, 31, 34, 42, 189n27, 192n61; Pavesich and, 41–44, 136, 206n32; peeping and, 26; philosophers and, 28, 53, 189n33; Poe on, 35–36; politics and, 51–52, 56; psychology and, 49; publicity and, 11, 34, 40–41, 47, 50, 191n59; public kissing and, 158; right to privacy and, 27, 35, 38, 41–42, 46, 53,

187n18, 188n24, 192n73; secrets and, 26–29, 32, 34, 38–50, 54–55; self-knowledge and, 31, 41, 46, 48, 192n61, 192n71; snapshot, 18, 26, 34–35, 38, 45, 52; subjectivity and, 107, 189n33; surveillance and, 44, 56–57; Svenson and, 23–27, 34, 56–57; technology and, 25, 27, 34–39, 44, 46, 50–51, 191n53; as theft, 45–54; Valéry and, 38; violations and, 26, 29, 39, 42–54, 56, 192n61; Warren and, 27, 34–35, 38–41, 46–53, 187n18, 189n25, 190n34, 190n42, 191n53, 192n61, 192n73

Plato, 16, 204n12, 213n54

pluralism, 81, 152–153, 163, 174, 211n20

plutocracy, 139–140

Poe, Edgar Allen, 35–36, 44, 97

Poetics of Relation, The (Glissant), 143–144

poetry, 80, 120

political economy, 2, 11, 28, 181n6

politics: agency and, 15, 74, 81–82; Arendt and, 211n20; Barthes on, 52; capitalism and, 5–7, 139, 181n6, 198n36; ethics and, 2–13, 51, 56, 92, 122, 143–144, 163–164, 174, 177–178, 181n6, 182n7; Foucault and, 53; hiding and, 89, 92, 97, 116; human depth and, 150–157, 163–167, 174, 211n20, 213n48; liberal philosophy and, 28; memory and, 119–122, 127–130, 136, 139, 143–144, 207n43; moral issues and, 37, 188n24, 200n23; personal appearance and, 33; photography and, 51–52, 56; selfhood and, 209n69; unaccountability and, 74, 81–82; Zuboff and, 181n6

Principles of Psychology, The (James), 122

privacy: biographical dimension and, 66–67, 70, 72, 76, 194n11; historical perspective on, 28–33; moral issues and, 58–64, 69, 73–77, 80–81, 84; sacredness of, 27, 29–34, 48, 126, 154, 156–157, 187n16, 192n61; spontaneity and, 31–32, 35, 52, 171, 187n16; spying and, 17, 39, 58–64, 67, 70, 79, 194n8, 194n11, 195n16

privacy settings, 5, 7, 177

propositional knowledge, 47, 156, 183n13, 198n42

Proust, Marcel, 165

psychology: agency and, 63, 74, 77, 79, 86; Freudian, 99, 120, 200n22; hiding and, 93, 99–104, 109, 112; human depth and, 148, 172, 214n66; James and, 122; memory and, 118, 120, 122, 135; photography and, 49; privacy and, 3, 19–20; self-reflection and, 3, 74–75, 170; tethering and, 20

public good: Arendt and, 151; human depth and, 21–22, 151, 156–167; right to be forgotten and, 136

publicity: agency and, 72–75; Arendt on, 151–152; Brandeis and, 90, 154; criminal branding and, 130–137; documentation and, 17, 21, 119, 136–144, 175, 183n16, 194n13; fixity and, 137–144; hiding and, 90–91, 105, 111, 114–116, 199n10; human depth and, 151–159, 166, 175; memory and, 120; photography and, 11, 34, 40–41, 47, 50, 191n59; Ruskin on, 89–90; use of term, 115–116; Warren and, 90, 154

Putnam, Robert, 114

Quashie, Kevin, 105–106

Rachels, James, 84–86

racism, 37, 82, 128, 183n18, 193n3, 200n23

radio: hiding and, 91, 93, 103, 111, 199n10; memory and, 125–126; privacy and, 11, 17

Rear Window (film), 25

Reiman, Jeffrey, 112–113, 116, 202n50

Ribot, Théodule-Armand, 122

Rieff, David, 128

right to be forgotten: debates over, 11; Google and, 205n23; human depth and, 165, 171, 205n23, 205n27, 206n36, 207n43, 214n64; memory and, 21, 118–119, 127–139, 143, 205n23, 205n27, 206n36; oblivion and, 11, 16, 21

right to privacy: agency and, 60–62, 84; Cooley and, 212n32; hiding and, 89;

right to privacy (*continued*)
  human depth and, 158; memory and,
  129–130, 205n27; moral issues and, 9, 38,
  41–43, 53, 60–62, 84, 129, 182n7, 187n18,
  188n24; Nissenbaum and, 182n9; photography and, 27, 35, 38, 41–42, 46, 53, 187n18,
  188n24, 192n73; Rothman and, 9–10
"Right to Privacy, The" (Warren and
  Brandeis), 27, 34–35, 38–39, 41, 46–51, 53,
  90, 154, 187n18, 188n24, 189n25, 189n27,
  191n53, 194n9
Romanticism, 30, 52, 68, 90, 187n18
Rosen, Michael, 193n6
Rosenblum, Nancy, 47
Rothman, Joshua, 9–10, 71, 166
Ruskin, John, 89–90

"Sacred Privacy of Home, The" (article), 29,
  48, 187n16
Scarlet Letter, The (Hawthorne), 29
Schiller, Friedrich, 46, 187n18
Schopenhauer, Arthur, 33, 45
scrutiny, 36, 45, 52, 79, 141
secrets: agency and, 58–61, 71, 80, 82, 85,
  194n8; Bok on, 184n19; confidentiality
  and, 20, 184n26, 185n10, 191n59; hiding
  and, 94–100, 104–105, 110, 199n15,
  200n22; human depth and, 162–163, 166,
  170, 172, 175; invasive questions and,
  54–57; memory and, 120, 144; opacity
  and, 13; personality and, 32, 47–48, 144;
  photography and, 26–29, 32, 34, 38–50,
  54–55; privacy and, 3, 12–16, 20, 22,
  184n19; *Voyeur's Motel* and, 71, 80, 85,
  197n34
self, the, 3, 22, 30–33, 44–47, 52–54, 66,
  68–70, 73–80, 83, 86–87, 89, 101–104,
  106–110, 112–113, 116, 136–137, 142–149, 153,
  155, 161–164, 170–172, 175–176
self-governance, 2, 68, 80
self-knowledge: agency and, 74–79, 87,
  195n17; Apollonian temple and, 2;
  boundaries and, 168; hiding and, 102;

human depth and, 149, 157, 160, 162, 165,
  168–171; limits of, 2–3, 9, 22, 35, 41, 48,
  78–79, 87, 126, 144, 161–163, 168–172,
  192n71; memory and, 125, 142–146;
  oblivion and, 2, 19, 22, 79–80, 118–176;
  photography and, 31, 41, 46, 48, 192n61,
  192n71; shallowness and, 2, 22, 147, 150,
  158–161, 164, 169, 176, 212n38
self-reflection, 3, 74–75, 170
self-relation: agency and, 15, 19, 80, 197n33;
  Arendt and, 155–157; hiding and,
  104–106; human depth and, 155–157, 170;
  modes of, 104–106; self-ignorance and,
  192n71
self-trust: Arendt and, 173–174; coming apart
  and, 171; darkness and, 172–173; Emerson
  and, 171; hiding and, 102; human depth
  and, 9, 102, 167–172, 175; Kupfer and,
  173–174; oblivion and, 9; story of K and,
  171–172; surveillance and, 172–173
Shakespeare, William, 16, 99, 157
sleep, 44, 64; Borges and, 124, 150; Darrieussecq on, 210n74; death and, 149–150;
  Diski on, 148–149; distraction and,
  204n13; dreams and, 122, 148–150, 155,
  181n3, 188n24; health data on, 1; human
  depth and, 148–150, 172–173; insomnia,
  64, 123, 126, 145–148, 210n74; oblivion
  and, 126, 145, 148–150, 172; phone's
  interrupting, 111; tethered addictions
  and, 100; voyeurism and, 68, 77
Small, Scott, 123
smartphones: GPS and, 172; health apps
  and, 120; hiding and, 88–89, 93, 95,
  105–111, 114, 202n45; technology and, 12,
  17, 88–89, 93, 95, 105–111, 114, 120, 139, 172,
  202n45; text messages and, 12, 88–89;
  Turkle on, 202n45
Smith, Justin E. H., 89
snapshots, 18, 26, 34–35, 38, 45, 52
Snowden, Edward, 58
social media: connectivity and, 11, 92, 103;
  Facebook, 7–8; hiding and, 89, 92, 103,

107–110, 202n41; human depth and, 164; informational privacy and, 7; memory and, 142; oversharing and, 92; privacy settings and, 5; public arena and, 19; Turkle on, 89, 107–109, 202n45

social ontology, 13, 49, 112, 151–167

Socrates, 207n39

solitude, 30, 82–83, 93, 102–106, 109, 111, 116, 162, 166

spontaneity, 31–32, 35, 52, 171, 187n16

spying: biographical dimension and, 66–67, 70, 72, 76, 194n11; consent and, 5, 11, 25, 45, 56, 72, 91, 105, 108, 195n16; Foos and, 19, 58–64, 67, 70, 79, 194n8, 195n16; neighbors and, 17, 39

stalking, 108, 111, 120, 139

Stein, Gertrude, 91, 199n10

stereotypes, 31

Stevens, Benjamin, 41

Stevens, Wallace, 91, 93, 124–125, 149

Stewart, James, 25

subjectivity: agency and, 196n20; Foucault and, 53; habits of perception and, 37; hiding and, 89, 107, 112, 114; human depth and, 159, 164; media and, 89; photography and, 107, 189n33

surveillance: agency and, 58–60, 83; capitalism and, 5, 139, 181n6, 198n36; CCTV, 19; citizen, 4, 8; GPS, 19, 172; hiding and, 91–92, 104; human depth and, 164, 172, 214n65; memory and, 136, 139–140; NSA and, 58; photography and, 44, 56–57; privacy and, 4–8, 13, 18–20; right to be forgotten, 11, 16, 21, 118–119, 127–139, 143, 165, 171, 205n27, 206n36, 207n43, 214n64; self-trust and, 172–173; technology and, 4, 11, 18, 44, 164, 172, 214n65

Svenson, Arne, 139; First Amendment and, 25; *Foster v. Svenson*, 184n23, 185n1, 188n24; moral issues and, 23–27; *The Neighbors*, 23–25, 56–57; New York State Court of Appeals and,

26–27; photography and, 23–27, 34, 56–57

Sweeney, Latanya, 139

Talese, Gay, 19, 58

Taylor, Charles, 75, 86, 170, 209n69

technology: addiction to, 109, 114; agency and, 63, 66, 85; disruptive, 18; ethics and, 6, 8, 10–11, 142–144, 160, 164, 177–178; facial-recognition, 11, 18–19; health apps, 120; hiding and, 89–90, 103, 107–110, 113–114, 199n15; human depth and, 149, 151, 160–161, 164, 172–173; innovation and, 10–11, 31, 190; Internet and, 7–11 (*see also* Internet); as invasion, 18, 25, 27, 34–35, 39, 46, 50, 90; memory and, 119–120, 131–132, 135, 137, 141–144; perception and, 66; photography and, 25, 27, 34–39, 44, 46, 50–51, 191n53; radio, 11, 17, 91, 93, 103, 111, 125–126, 199n10; smartphones, 12, 17, 88–89, 93, 95, 105–111, 114, 120, 139, 172, 202n45; solutions and, 173, 177–178; surveillance and, 4, 11, 18, 44, 164, 172, 214n65; television, 11, 59, 91, 103, 111, 199n10; unaccountability and, 63, 66, 85

technophobia, 4

television, 11, 59, 91, 103, 111, 199n10

tethering: audience and, 110; connectivity and, 100–109; emotion and, 93; experience and, 100–104; Freitas and, 109–110; hiding and, 93, 100–110, 113, 116–117; memory and, 118, 131, 143, 146; psychological orientation and, 20; self and, 106–109; Turkle on, 107–109

text messaging, 12, 88–89

Thirty Years' War, 16–17

totalitarianism, 93, 116, 151

Tractatus Logico-Philosophicus (Wittgenstein), 163

Treaty of Westphalia, 16–17

trespass: Arendt on, 116; harmless, 194n8; invasion and, 48; onto property, 29, 107; undiscovered, 64–65; unlawful entry, 26

trustworthiness: Arendt and, 173–174;
    coming apart and, 171; darkness and,
    172–173; human depth and, 167–176;
    Kupfer and, 173–174; problem-solving
    and, 167–170; self-trust and, 9, 102,
    167–172, 175
Turkle, Sherry, 89, 107–109, 202n45

unaccountability: agency and, 71–72, 76–87,
    195n16; human depth and, 162, 174;
    identity and, 69, 74; intimacies and,
    83–87; mapping, 80–83; memory and,
    118, 145; oblivion and, 8, 19; politics and,
    74, 81–82; protection of, 8; technology
    and, 63, 66, 85
unuttered speech, 51

Valéry, Paul, 38, 107
Véliz, Carissa, 140
Victorian era, 28, 31, 34, 50–51
violation: access and, 14; accountability and,
    19, 71, 77–83, 86, 118, 174, 196n20; agency
    and, 58–72; assault, 64–66; biographical
    dimension and, 66–67, 70, 72, 76, 194n11;
    burglars and, 26; consent and, 5, 11, 25,
    45, 56, 72, 91, 105, 108, 195n16; Foos and,
    19, 58–73, 76–80, 95, 193n4, 195n16,
    197n34; hiding and, 97, 104–105, 111;
    human depth and, 158; memory and,
    129–131, 205n28; NSA and, 58; person-
    ality and, 42–54, 187n18; photography
    and, 26, 29, 39, 42–54, 56, 192n61; spying
    and, 17, 39, 58–64, 67, 70, 79, 194n8,
    194n11, 195n16; stalking and, 108, 111, 120,
    139; undiscovered assault, 64; well-being
    and, 61–64
violence, 17, 81, 128

voyeurism: accountability and, 71–83;
    autonomy and, 63, 74, 77, 81, 84;
    biographical dimension, 66–67, 70, 72,
    76, 194n11; Foos and, 19, 58–73, 76–80, 95,
    193n4, 195n16, 197n34; hackers, 63–66;
    harm and, 19, 58–69, 72, 81, 193n4;
    Manor House Motel and, 58–59; moral
    issues and, 59–63, 77, 138; sleep and, 68,
    77; spying and, 19, 58–64, 67, 70, 79,
    194n8, 195n16; stalking and, 108, 111, 120,
    139; well-being and, 61–64
Voyeur's Motel (Talese): forgetting and, 138;
    invasion and, 19, 58, 71; moral issues and,
    193n3, 196n16; peeping and, 19, 60–69,
    72, 76–77; secrets and, 71, 80, 85, 197n34;
    violations and, 58–72

Warren, Samuel, 27, 34–35, 38–41, 46–53, 90,
    154, 187n18, 189n25, 190n34, 190n42,
    191n53, 192n61, 192n73, 194n9; newspa-
    pers and, 90; on property, 48–49;
    photography and, 27, 34–35, 38–41,
    46–53, 187n18, 189n25, 190n34, 190n42,
    191n53, 192n61, 192n73; publicity and, 90,
    154; "The Right to Privacy" and, see
    "Right to Privacy, The"; spirituality and,
    48, 194n9
Whitman, Walt, 191n56
Wittgenstein, Ludwig, 163, 211n25
Woolf, Virginia, 9–10, 73, 166

X-rays, 26, 34, 36

yachting and madness, 130

Zuboff, Shoshana, 181n6
Zuckerberg, Mark, 8